Cracking the Finance Quant Interview

Jean Peyre, Cracking the Finance Quant Interview.

Paperback Edition April 2024
For any question, remark or typo please email admin@editionsducourt.com

# Cracking the Finance Quant Interview

*Jean Peyre*

Éditions Ducourt

# Contents

# Chapter 1

# Brainteasers

# Brainteasers

Difficulty: ♠ Medium ♠♠ Hard ♠♠♠ Very Hard

## 1.1 Triangle Impossible I ♠ (Goldman Sachs)

We randomly break a stick of length 1 into three pieces, What is the probability that the pieces can form a triangle? (breaking points are uniformly distributed between 0 and 1).

---

Solution in page 13

## 1.2 Triangle Impossible II ♠♠ (Goldman Sachs)

Let A, B and C three independent random variables uniformly distributed between 0 and 1. We make 3 sticks respectively of length A, B and C. What is the probability that the sticks can form a triangle?

---

Solution in page 14

## 1.3 A Prime Number ♠ (Commerzbank)

We consider a prime number $p \geq 5$. Prove that 24 divides $(p^2 - 1)$ i.e.
$24|(p^2 - 1)$.

---

Solution in page 16

## 1.4 The Last Prime ♠ (Goldman Sachs)

Prove that there is an infinity of prime numbers.

---

Solution in page 16

## 1.5 Erdős Subsequences ♠♠♠ (Goldman Sachs)

We consider a sequence $u_n$, $n \in [1, 300]$ composed of distinct real numbers. Show that we can extract a strictly increasing or strictly decreasing subsequence $u_{\phi(n)}$ containing at least 17 elements.

Solution in page 16

## 1.6 Omelette ♠♠ (Google)

You are given two eggs, and access to a 100-storey building. Both eggs are identical. The aim is to find out the highest floor from which an egg will not break when dropped out of a window from that floor. If an egg is dropped and does not break, it is undamaged and can be dropped again. However, once an egg is broken, that's it for that egg. If an egg breaks when dropped from floor n, then it would also have broken from any floor above that. If an egg survives a fall, then it will survive any fall shorter than that.

What strategy should you adopt to minimize the number of egg drops it takes to find the solution?

Solution in page 17

## 1.7 A Hard Pill to Swallow ♠ (HSBC)

A blind man is alone on a deserted island. He has two blue pills and two red pills. He must take exactly one red pill and one blue pill to survive. How does he do it?

Solution in page 18

## 1.8 Game Theory ♠♠♠ (Goldman Sachs)

Player A invites player B to play the following game: A picks an integer $n$ between 1 and 100, and writes it on a paper. B tries to guess $n$. If he succeeds, he receives $n$ dollars. What is the fair price of the game, and what should be the strategy of B?

Solution in page 19

## 1.9 Stable Equilibrium I ♠ (Goldman Sachs)

N tigers circle around an antelope. If a tiger eats an antelope or another tiger, it falls asleep and it becomes a potential meal for the remaining tigers. Tigers will eat if it does not endanger their life. The antelope keeps grazing quietly. Why?

Solution in page 19

## 1.10 Stable Equilibrium II ♠♠ (Goldman Sachs)

100 silent monks live in a monastery with no mirrors or reflective surfaces and one important rule: no red eyes! If a monk discovers he has red eyes he commits suicide at midnight. They live happily together in peace until a tourist visiting the monastery says "at least one of you has red eyes!". What happens next?

---

Solution in page 20

## 1.11 Need For Speed ♠ (Goldman Sachs)

A car travels 100km in 1 hour. Show that, at some point, its speed was exactly 100km/h.

---

Solution in page 21

## 1.12 Russian Coin I ♠ (CitiBank)

Three players A, B and C sit around a table. They have a fair coin which gives heads or tails with a probability $\frac{1}{2}$. Player A tosses the coin, if he gets heads he wins, and the game is over. Otherwise he gives the coin to B, who is sitting at his right hand side. If B gets heads he wins, otherwise he gives the coin to C etc... What is the probability for each player to win the game?

---

Solution in page 21

## 1.13 Russian Coin II ♠ (CitiBank)

Three players A, B and C sit around a table. They have a strange coin which gives heads or tails with a probability $\frac{1}{4}$, and stays stuck on its side with a probability $\frac{1}{2}$. Player A tosses the coin, if he gets side he wins, and the game is over. Otherwise if A gets heads he gives the coin to B, who is sitting at his right hand side. If A gets tails he gives the coin to C, who is sitting at his left hand side. The next player restarts the same process. What is the probability for each player to win the game?

---

Solution in page 22

## 1.14 Be My Guest ♠♠ (Goldman Sachs)

$N$ guests are queuing at the entrance to get seated at a wedding table. Every guest has an assigned seat number but the first guest to choose his seat is too drunk and takes a random seat. The remaining guests choose their seat according to the following rule:

- if their assigned seat is available they take it
- if their assigned seat is taken they choose randomly an available seat

What is the probability that the last person gets his assigned seat?

---

Solution in page 23

## 1.15 4 Coins, 1 Table ♠♠♠ (CitiBank)

4 coins are placed at the corners of a rotating table and the player is blindfolded. At every turn, the player can flip as many coins as he wants, and ask the game master if the coins are all showing heads. If they are all heads, the players wins, otherwise the game master can arbitrarily rotate the table before the next turn. Is there a winning strategy for the player?

---

Solution in page 24

## 1.16 N Coins, 1 Table ♠♠♠ (UBS)

2 players take turns placing coins on a large perfectly round table. Coins can not overlap and all the coin surface must be in contact with the table. The first player who can't place a coin loses. Is it better to play first and is there a winning strategy?

---

Solution in page 25

## 1.17 Regression Mirror ♠ (GResearch Quant Sample)

Suppose that $X$ and $Y$ are mean zero, unit variance random variables. If least squares regression (without intercept) of $Y$ against $X$ gives a slope of $\beta$ (i.e. it minimises $\mathbb{E}[(Y - \beta X)^2]$), what is the slope of the regression of X against Y?

---

Solution in page 26

## 1.18 Bayesian Kids ♠ (GResearch Quant Sample)

I meet someone with 2 children, and I learn that one of the children is a boy. What's the probability that the other child is also a boy? What if one of the children is a boy born on a Tuesday?

Solution in page 26

## 1.19 The Last Digit ♠ (GResearch Quant Sample)

Consider all 100 digit numbers, i.e. those between 0 to $(10^{100} - 1)$, inclusive. For each number, take the product of non-zero digits (treat the product of digits of 0 as 1), and sum across all the numbers. What's the last digit?

Solution in page 27

## 1.20 Repeated Contraction ♠ (GResearch Quant Sample)

Let $R(n)$ be a random draw of integers between 0 and $n-1$ (inclusive). I repeatedly apply $R$, starting at $10^{100}$. What's the expected number of repeated applications until I get zero?

Solution in page 28

## 1.21 Domino's Pizza ♠♠ (GResearch Quant Sample)

How many ways are there to tile dominos (with size $2 \times 1$) on a grid of $2 \times$ n? How about on a grid of $3 \times 2$n?

Solution in page 28

## 1.22 Nash's Car ♠♠♠ (GResearch Quant Sample)

A company has a competition to win a car. Each contestant needs to pick a positive integer. If there's at least one unique choice, the person who made the smallest unique choice wins the car. If there are no unique choices, the company keeps the car and there's no repeat of the competition. It turns out that there are only three contestants, and you're one of them. Everyone knows before picking their numbers that there are only three contestants. How should you make your choice?

Solution in page 32

## 1.23 Correlation Impossible I ♠ (UBS)

If X, Y and Z are three random variables such that X and Y have a correlation of 0.9, and Y and Z have correlation of 0.8, what are the minimum and maximum correlation that X and Z can have?

Solution in page 33

## 1.24 Correlation Impossible II ♠♠ (UBS)

If $X_1, X_2 ... X_n$ are n random variables such that

$$\text{Corr}(X_i, X_j) = \rho \quad \text{for } i \neq j$$

what are the minimum and maximum values that $\rho$ can have?

Solution in page 34

## 1.25 The Dark Side of the Die ♠ (Societe Generale)

How many times do I have to roll a die until all six sides appear?

Solution in page 35

## 1.26 Bonus Day ♠ (UBS)

Five pirates $P_i$ have 100 gold coins. They have to divide up the loot. In order of seniority (suppose pirate $P_5$ is most senior, $P_1$ is least senior), the most senior pirate proposes a distribution of the loot. They vote and if at least 50% accept the proposal, the loot is divided as proposed. Otherwise the most senior pirate is executed, and they start over again with the next senior pirate. Which solution does the most senior pirate propose? Assume they are very intelligent and extremely greedy (and that they would prefer not to die).

Solution in page 36

## 1.27 Secret Polynomial ♠♠ (BNP)

We consider a polynomial $P(x)$ of which all coefficients are positive integers ($a_i \geq 0$). The polynomial is in a black box and we can only retrieve its value in given points. In how many points do we need to value the polynomial in order to find the values of all the coefficients?

Solution in page 37

## 1.28 Drunk Mutant Ninja Ant ♠♠ (Nomura)

An ant starts a walk from a cube vertex, it walks on the edges and at every vertex it chooses to walk one of the available edges (including the edge it came from) with an equal probability. How many edges will the ant cross in average to come back to the starting point?

Solution in page 37

## 1.29 Dog Day Afternoon ♠♠ (Deutsche Bank)

You are standing at the centre of a circular field of radius R. The field has a low wire fence around it. Attached to the wire fence (and restricted to running around the perimeter) is a large, sharp-fanged, hungry dog. You can run at speed $v$, while the dog can run four times as fast. What is your running strategy to escape the field?

Solution in page 38

## Chapter 2

# Brainteasers - Solutions

# Brainteasers - Solutions

## 2.1 Triangle Impossible I - Solution

**Question :** We randomly break a stick of length 1 into three pieces, What is the probability that the pieces can form a triangle? (breaking points are uniformly distributed between 0 and 1).

**Solution :** An elegant and effective method to solve the problem is to visualize it geometrically. Let us define $x$ and $y$ the two breaking points and let us assume $x \geq y$ (the case $x \leq y$ is symmetric). In order to form a triangle we need the longest piece to be shorter than the sum of the two other pieces.

$$\text{In the case } x \geq y \text{ we need} \begin{cases} x \geq \frac{1}{2} \\ y \leq \frac{1}{2} \\ (x-y) \leq \frac{1}{2} \end{cases} \tag{1}$$

We translate these conditions in the chart below, the chart on the left side is for the case $x \geq y$ and the chart on the right side is for the general case. The grey area represents the cases where we can form a triangle, we see that in both charts the grey area is a quarter of the full area. The probability to form a triangle is therefore $\frac{1}{4}$.

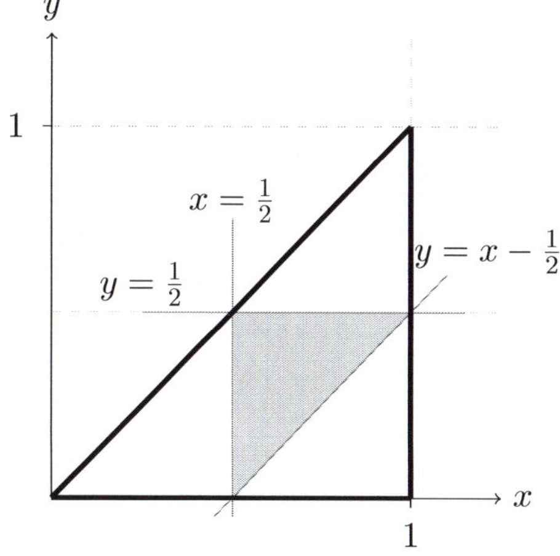


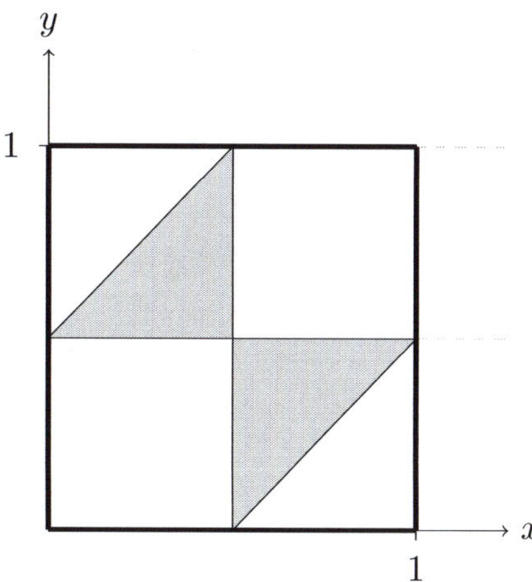

We can also answer using integrals. We consider here that $x \leq \frac{1}{2}$ (the other case is symmetric), and for $x \in [0, \frac{1}{2}]$ we see that $y$ must be greater than $\frac{1}{2}$ and lower than $(x + \frac{1}{2})$.

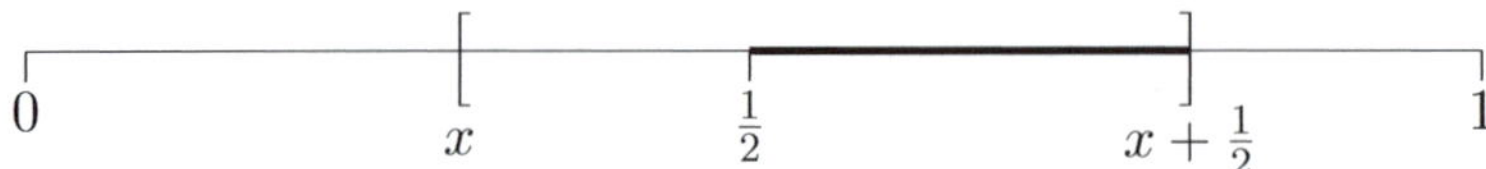


Translated into integrals, and using the symmetry argument we have $P$ the probability to form a triangle

$$P = 2\int_0^{\frac{1}{2}} x dx = 2\left[\frac{x^2}{2}\right]_0^{\frac{1}{2}} = \frac{1}{4}$$

## 2.2 Triangle Impossible II - Solution

**Question :** Let A, B and C three independent random variables uniformly distributed between 0 and 1. We make 3 sticks respectively of length A, B and C. What is the probability that the sticks can form a triangle?

**Solution :** This problem can be solved elegantly with a drawing. We work this time on a cube and the condition to form a triangle is that no stick is longer that the sum of the others.

$$\text{we need}\begin{cases} A \leq (B+C) \\ B \leq (A+C) \\ C \leq (B+A) \end{cases} \tag{2}$$

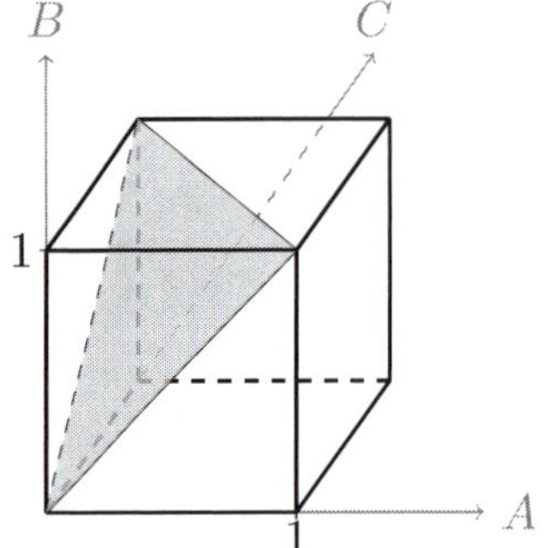


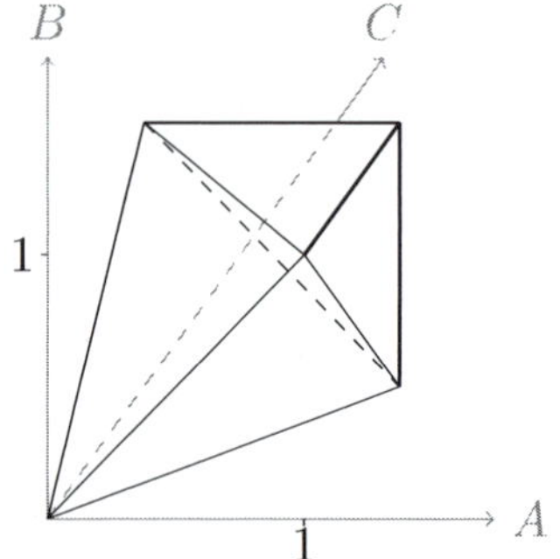

The valid volume is the diamond on the right. To calculate this volume we subtract 3 times the volume of the pyramid below from the the original cube.

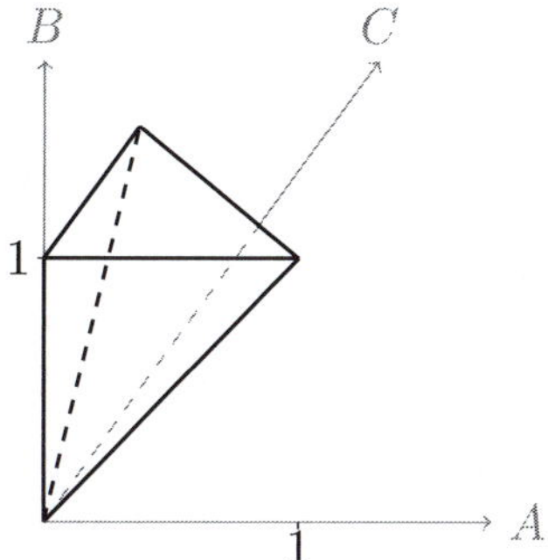


The volume of this pyramid is $V_{pyramid}$ = base.height.$\frac{1}{3} = \frac{1}{2}.1.\frac{1}{3} = \frac{1}{6}$. Therefore $V_{diamond} = 1 - 3.\frac{1}{6} = \frac{1}{2}$. The probability we are looking for is $P = \frac{1}{2}$.

The question can also be solved with integrals

$$I = \int_{x=0}^{x=1} \int_{y=x}^{y=1} \int_{z=y}^{1\wedge(x+y)} dxdydz$$

$$I = \int_{x=0}^{x=1} \int_{y=x}^{y=1} \Big[x \wedge (1-y)\Big] dxdy$$

$$I = \int_{x=0}^{x=1} \int_{Y=0}^{Y=1-x} \Big[x \wedge Y\Big] dxdY; \quad Y = (1-y)$$

We decompose $I$ into 3 terms

$$I = I_1 + I_2 + I_3$$

$$I = \int_{x=0}^{x=\frac{1}{2}} \int_{Y=0}^{Y=x} YdxdY + \int_{x=0}^{x=\frac{1}{2}} \int_{Y=x}^{Y=1-x} xdxdY + \int_{x=\frac{1}{2}}^{x=1} \int_{Y=0}^{Y=1-x} YdxdY$$

$$I_1 = \int_{x=0}^{x=\frac{1}{2}} \int_{Y=0}^{Y=x} YdxdY = \int_{x=0}^{x=\frac{1}{2}} \frac{x^2}{2} dx$$

$$I_2 = \int_{x=0}^{x=\frac{1}{2}} \int_{Y=x}^{Y=1-x} xdxdY = \int_{x=0}^{x=\frac{1}{2}} x(1-2x)dx$$

$$I_3 = \int_{x=\frac{1}{2}}^{x=1} \int_{Y=0}^{Y=1-x} YdxdY = \int_{x=\frac{1}{2}}^{x=1} \frac{(1-x)^2}{2} dx = \int_{s=\frac{1}{2}}^{s=0} -\frac{s^2}{2} ds = \int_{x=0}^{x=\frac{1}{2}} \frac{s^2}{2} dx$$

$$I = \int_{x=0}^{x=\frac{1}{2}} \frac{x^2}{2} + \frac{x^2}{2} + x(1-2x)dx = \int_{x=0}^{x=\frac{1}{2}} x(1-x)dx = \left[\frac{x^2}{2} - \frac{x^3}{3}\right]_0^{\frac{1}{2}} = \frac{1}{12}$$

$$P = 6I = \frac{1}{2}$$

## 2.3 A Prime Number - Solution

**Question :** We consider a prime number $p \geq 5$. Prove that 24 divides $(p^2 - 1)$ i.e. $24|(p^2 - 1)$.

**Solution :** In order to prove that $24|(p^2 - 1)$ we can prove that $8|(p^2 - 1)$ and $3|(p^2 - 1)$.

We note first that $p^2 - 1 = (p - 1)(p + 1)$, and $p$ being a prime number, $(p - 1)$ and $(p + 1)$ are two consecutive even integers. Therefore both can be divided by 2 and one of them can be divided by 4. We have proved that $8|(p^2 - 1)$.

We also observe that $p - 1$, $p$ and $p + 1$ are 3 consecutive integers. One of them is necessarily divisible by 3 and it can not be $p$ because it is a prime number. This shows that $3|(p^2 - 1)$ and therefore $24|(p^2 - 1)$.

## 2.4 The Last Prime - Solution

**Question :** Prove that there is an infinity of prime numbers.

**Solution :** This is a classic proof in number theory. We proceed by contradiction, we assume that the set of all prime numbers is finite $\{n_0, n_1, ..., n_M\}$. We consider now the integer

$$K = 1 + \prod_{i=0}^{M} n_i$$

$K$ can not be divided by any of the prime numbers in our set because for all of them we have $K$ modulo $n_i$ equal to 1 i.e. $K[n_i] = 1$. This means that $K$ is a prime number which was not in our set, hence the contradiction.

## 2.5 Erdős Subsequences - Solution

**Question :** We consider a sequence $u_n$, $n \in [1, 300]$ composed of distinct real numbers. Show that we can extract a strictly increasing or strictly decreasing subsequence $u_{\phi(n)}$ containing at least 17 elements.

**Solution :** The question is about the longest monotonous subsequence that can be extracted from a given sequence. Intuitively this length increases with the length of the original sequence. We denote $I_i$ (resp. $D_i$) the longest increasing (resp. decreasing) subsequence which last element is $u_i$. The application $i \mapsto \{I_i, D_i\}$ is injective

$$m < n \Rightarrow \begin{cases} u_m < u_n; I_n > I_m \\ \text{or} \\ u_m > u_n; D_n > D_m \end{cases}$$

Therefore once $n > p^2$ at best we can fill the square $[1,p]\text{x}[1,p]$ and be guaranteed to find a monotonous subsequence of length $p+1$. Actually for any $n$ we can find a monotonous subsequence of length $\lceil\sqrt{n}\rceil$. In our case $n = 300$ and we can extract 17 ordered elements.

So why is Erdős in the title? Because the Erdős–Szekeres theorem guarantees that any sequence of distinct real numbers with length at least $(r-1)(s-1)+1$ contains a monotonically increasing subsequence of length $r$ or a monotonically decreasing subsequence of length $s$. In this case $r = s = 17$ and 290 is the required number of elements in the sequence.

## 2.6 **Omelette** - Solution

**Question :** You are given two eggs, and access to a 100-storey building. Both eggs are identical. The aim is to find out the highest floor from which an egg will not break when dropped out of a window from that floor. If an egg is dropped and does not break, it is undamaged and can be dropped again. However, once an egg is broken, that's it for that egg. If an egg breaks when dropped from floor n, then it would also have broken from any floor above that. If an egg survives a fall, then it will survive any fall shorter than that.

What strategy should you adopt to minimize the number of egg drops it takes to find the solution?

**Solution :** The objective is to minimize the number of attempts in the worst case. If we had only one egg to solve the problem we would have needed to start at the first floor and to go up one floor for every new attempt. In the worst case we would have needed 100 attempts. If we have 2 eggs we can improve this strategy and skip floors when using the first egg. If the first egg breaks we can single out an interval of floors. We can then use the second egg to test the floors in the interval one by one from the bottom. The crucial question is the choice of intervals to skip with the first egg. We denote $u_i$ the sequence of floors from which the first egg is thrown and $W(k)$ the number of attempts needed in the worst case to solve the problem with 2 eggs and $k$ floors. After the first attempt at $u_1$, if the egg breaks we need try all the floors between 1 and $u_i - 1$. Otherwise we still have 2 eggs and

$(100 - u_i)$ remaining floors to test

$$W(100) = \max\left(u_1, 1 + W(100 - u_1)\right)$$

we repeat the same reasoning until the $i^{th}$ floor

$$W(100) = \max\left(u_1, u_2 - u_1 + 1, u_3 - u_2 + 2, \ldots, 1 + W(100 - u_i)\right)$$

we denote $v_i$ the increments sequence $v_i = u_i - u_{i-1}$, and $v_1 = u_1$. The equation becomes

$$W(100) = \max\left(v_1, v2 + 1, v_3 + 3, \ldots, 1 + W\left(100 - \sum_{k=1}^{i} v_k\right)\right)$$

and the full formula for the number of attempts is

$$W(100) = \max\left(v_1, v2 + 1, v_3 + 2, \ldots, v_n + n - 1\right)$$

we minimize this maximum when all the arguments are equal. On the other hand the increments sum to 100

$$\sum_{k=1}^{n} v_k = 100$$

We denote $M = v_i + i - 1$, and for a given $n$ the condition on $M$ is

$$\sum_{k=1}^{n} (M - k + 1) = nM + n - \sum_{k=1}^{n} k > 100$$

$$M > \frac{100}{n} - 1 + \frac{n+1}{2}$$

we calculate the derivative of the right side to find that the minimum is reached for $n = \sqrt{200} \approx 14.14$ and $M = 13.642$. In the worst case we will need $M = 14$ attempts and the sequence of floors to test with the first egg is

$$14, 27, 39, 50, 60, 69, 77, 84, 90, 95, 99, 100$$

## 2.7 A Hard Pill to Swallow - Solution

**Question :** A blind man is alone on a deserted island. He has two blue pills and two red pills. He must take exactly one red pill and one blue pill to survive. How does he do it?

**Solution :** Break each of the pills in half, as you do this pop one half in your mouth and discard the other half.

## 2.8 Game Theory - Solution

**Question :** Player A invites player B to play the following game: A picks an integer $n$ between 1 and 100, and writes it on a paper. B tries to guess $n$. If he succeeds, he receives $n$ dollars. What is the fair price of the game, and what should be the strategy of B?

**Solution :** A tempting strategy for player A is to pick the lowest number 1 and to be guaranteed to lose at most 1. The expected loss of the winning strategy will have to be lower than 1. The key in this type of question is to observe that both players have access to the same amount of information. Therefore player B will guess the optimal strategy of A and take full advantage of it. We denote $p(i)$ the discrete probability distribution that A decides to use $i$ for his choice. When player B picks a number he has an expected gain equal to $g_i = p(i).i$. Remember that player B will guess the probability distribution $p$, he will try to maximize his gain and player A will minimize the quantity

$$M = \max_{i \in [1,100]} g_i$$

This maximum is minimized when all the elements are equal $p(i).i = \lambda$. We find $\lambda$ using the probability distribution properties

$$\sum_1^{100} p(i) = \sum_1^{100} \frac{\lambda}{i} = 1$$

$$\lambda = \frac{1}{\sum_1^{100} \frac{1}{i}} \approx \frac{1}{1 + \ln(n)}$$

therefore player A will pick the number $i$ with a probability $p(i) = \frac{\lambda}{i}$. The expected loss (gain) for A (B) is $G = p(i).i = \lambda$. The numerical application with 100 numbers gives $G \approx 0.18$.

## 2.9 Stable Equilibrium I - Solution

**Question :** N tigers circle around an antelope. If a tiger eats an antelope or another tiger, it falls asleep and it becomes a potential meal for the remaining tigers. Tigers will eat if it does not endanger their life. The antelope keeps grazing quietly. Why?

**Solution :** In this classic type of question an unexpected equilibrium appears in a system. The best way to understand it is to start with a small number of tigers.

- 1 tiger: the tiger clearly eats the antelope, he does not need to worry about sleeping after the meal.
- 2 tigers: if a tiger eats the antelope he gets eaten by the other tiger. Tigers know that and decide to stay still. The system with 2 tigers is a stable system.
- 3 tigers: tigers have read this book and know that the 2 tigers system is stable, one of them eats the antelope, falls asleep and becomes the pray in a stable 2 tigers system.
- 4 tigers: tigers know that the 3 tigers system is unstable and prefer not to eat the antelope, the 4 tigers system is stable.

It appears that systems with an even number of tigers are stable. The antelope is relaxed because she has counted the tigers and found an even number.

## 2.10 Stable Equilibrium II - Solution

**Question :** 100 silent monks live in a monastery with no mirrors or reflective surfaces and one important rule: no red eyes! If a monk discovers he has red eyes he commits suicide at midnight. They live happily together in peace until a tourist visiting the monastery says "at least one of you has red eyes!". What happens next?

**Solution :** This is a different version of a classic type of equilibrium puzzles. We start with the cases with a low number of red eyed monks (RE group).

- Zero RE monk and the tourist lied to them: on the first day all the monks think that they are RE because they cannot see anyone else in RE. At midnight they all commit suicide. This bad prank should not happen because the tourist is assumed to tell the truth.
- 1 RE monk: the RE monk cannot see anyone else in RE and commits suicide at midnight.
- 2 RE monks: RE monks think on the first day that there is only one RE monk and they can see it. But no one commits suicide on the first night. At that point they realize that they are in a system with 2 RE and they both commit suicide on the second night.
- 3 RE monks: RE monks think that they are in a system with 2 RE, but no one commits suicide on the second night, they realize that it is a 3 RE system and they all commit suicide on the third night.

The pattern is clear, in conclusion in a system with $j$ RE monks, all the RE monks commit suicide on the $j^{th}$ night.

## 2.11 Need For Speed - Solution

**Question :** A car travels 100km in 1 hour. Show that, at some point, its speed was exactly 100km/h.

**Solution :** This is a recurrent type of question based on the continuity of a function or its derivative. We denote $x(t)$ the position of the car at time $t$. $x(0) = 0$, $x(1) = 100$ and the mean value theorem proves that

$$\exists c \in [0, 1] : \; x'(c) = \frac{x(1) - x(0)}{1} = 100$$

Alternatively we can use the intermediate value theorem, if the average speed is 100, the speed cannot always be higher than 100, and it cannot always be lower than 100. There exists therefore a moment $t_h$ where the speed is greater or equal to 100 and a moment $t_l$ where the speed is lower or equal 100. Therefore $x'(t_h) \geq 100$ and $x'(t_l) \leq 100$ and $\exists c : x'(c) = 100$.

## 2.12 Russian Coin I - Solution

**Question :** Three players A, B and C sit around a table. They have a fair coin which gives heads or tails with a probability $\frac{1}{2}$. Player A tosses the coin, if he gets heads he wins, and the game is over. Otherwise he gives the coin to B, who is sitting at his right hand side. If B gets heads he wins, otherwise he gives the coin to C etc... What is the probability for each player to win the game?

**Solution :** There is an elegant way to solve this problem based on the symmetry of the players position. We define $p_A$ (resp. $p_B$, $p_C$) the probability that player A (resp. B, C) wins the game and $p$ as follows

$$p = P\left\{\text{Player who starts wins the game}\right\}$$

we see clearly that $p_A = p$. By symmetry, if player A misses his first toss player B finds himself in the position of starting the same game. Therefore

$$p_B = P\left\{\text{A misses the first toss} \cap \text{Player who starts wins the game}\right\} = \frac{p}{2}$$

$$p_C = \frac{p}{4}$$

Also the probability that no one wins is zero

$$P\{\text{No one wins}\} = \lim_{\infty} \frac{1}{2}^n = 0$$

and

$$p_A + p_B + p_C = p + \frac{p}{2} + \frac{p}{4} = 1$$

$$p = \frac{4}{7} = p_A;\ p_B = \frac{2}{7};\ p_C = \frac{1}{7}$$

The question can also be solved with series. We find that

$$p_A = \frac{1}{2} + \frac{1}{2} \cdot \frac{1}{2}^3 + \cdots + \frac{1}{2} \cdot \frac{1}{2}^{3i}$$

$$p_A = \frac{1}{2} \sum_{i=0}^{\infty} \frac{1}{8}^i = \frac{1}{2} \frac{1}{1 - \frac{1}{8}} = \frac{4}{7}$$

## 2.13 Russian Coin II - Solution

**Question :** Three players A, B and C sit around a table. They have a strange coin which gives heads or tails with a probability $\frac{1}{4}$, and stays stuck on its side with a probability $\frac{1}{2}$. Player A tosses the coin, if he gets side he wins, and the game is over. Otherwise if A gets heads he gives the coin to B, who is sitting at his right hand side. If A gets tails he gives the coin to C, who is sitting at his left hand side. The next player restarts the same process. What is the probability for each player to win the game?

**Solution :** We denote $P(i|j)$ the probability of the player $j$ winning a game started by the player $i$. The probability of no one winning is equal to zero

$$P\{\text{No one wins}\} = \lim_{\infty} \frac{1}{2}^n = 0$$

therefore

$$P(A|A) + P(B|A) + P(C|A) = 1$$

We can visualize the possible outcomes after A turn

$$\text{A plays} \begin{cases} \text{A gets heads with a probability } \frac{1}{4};\ \text{B wins with a probability } P(B|B) \\ \text{A gets tails with a probability } \frac{1}{4};\ \text{B wins with a probability } P(B|C) \\ \text{A gets side with a probability } \frac{1}{2};\ \text{A wins} \end{cases}$$

and

$$P(B|A) = \frac{1}{4} P(B|B) + \frac{1}{4} P(B|C)$$

and by symmetry $P(B|B) = P(A|A)$ and $P(B|C) = P(C|A)$ giving

$$P(B|A) = \frac{1}{4}P(A|A) + \frac{1}{4}P(C|A)$$

By symmetry we also have

$$P(B|A) = P(C|A)$$

so the system of equations is

$$P(A|A) + 2P(B|A) = 1$$
$$\frac{3}{4}P(B|A) = \frac{1}{4}P(A|A)$$

We find $P(A|A) = \frac{3}{5}$ and $P(B|A) = P(C|A) = \frac{1}{5}$.

## 2.14 Be My Guest - Solution

**Question :** $N$ guests are queuing at the entrance to get seated at a wedding table. Every guest has an assigned seat number but the first guest to choose his seat is too drunk and takes a random seat. The remaining guests choose their seat according to the following rule:

- if their assigned seat is available they take it
- if their assigned seat is taken they choose randomly an available seat

What is the probability that the last person gets his assigned seat?

**Solution :** Let us say the drunk person's seat is the number 1 and the last person's assigned seat is $n$. If at any moment a displaced person randomly chooses the seat number $n$, then the last person cannot get his assigned seat. But the critical remark is that, if at any time a displaced person randomly chooses the seat number 1, then the last person get his assigned seat. The chain of displaced guests is a cyclical permutation of the chain of assigned seats and choosing the seat 1 closes the cycle.

| Guest | k | 1 | i | j |
|---|---|---|---|---|
| Seat | 1 | i | j | k |

Therefore, as long as the seats 1 and $n$ are available, the entering guest $k$ has the following options

$$\begin{cases} p = \frac{1}{k} \text{ to pick 1, the last person is not displaced} \\ p = \frac{1}{k} \text{ to pick n, the last person is displaced} \\ p = \frac{k-2}{k} \text{ to pick another seat, the choice between 1 and n is postponed} \end{cases}$$

we can ignore how often the choice between 1 and $n$ is postponed, when it finally happens the probabilities to choose 1 or $n$ are equal. The probability that the last person gets his assigned seat is $\frac{1}{2}$.

## 2.15 4 Coins, 1 Table - Solution

**Question :** 4 coins are placed at the corners of a rotating table and the player is blindfolded. At every turn, the player can flip as many coins as he wants, and ask the game master if the coins are all showing heads. If they are all heads, the players wins, otherwise the game master can arbitrarily rotate the table before the next turn. Is there a winning strategy for the player?

**Solution :** We use the notation [h,h,h,t] for the current coins position, where h stands for heads and t for tails. We group the coins positions in classes which are stable by cyclical permutation. That means for example that [t,h,h,h], [h,t,h,h], [h,h,t,h] are grouped in the same class. Each time the player asks the game master if the current position is a winning position he can also flip all the coins and test the complementary position too. Therefore we can include the complementary sets in the classes, which means that [t,h,h,h], [h,t,t,t], [h,t,h,h], [t,h,t,t] etc... are in a same class.

We use the notation [f,o,o,o] to indicate which coins are flipped by the player, f stands for flipped and o indicates that the coins are not flipped. Similarly we group the player moves in classes which are stable by cyclical permutation.

$$\text{Position Classes}\begin{cases} p_1 : [\text{h,h,h,h}] \\ p_2 : [\text{t,h,h,h}] \\ p_3 : [\text{t,t,h,h}] \\ p_4 : [\text{t,h,t,h}] \end{cases} \quad \text{Transition Classes}\begin{cases} t_1 : [\text{f,f,f,f}] \\ t_2 : [\text{f,o,o,o}] \\ t_3 : [\text{f,f,o,o}] \\ t_4 : [\text{f,o,f,o}] \end{cases}$$

The transition $t_1$ is used at every step to check the complementary position. The position class $p_1$ is a winning position. We draw the following transition diagram

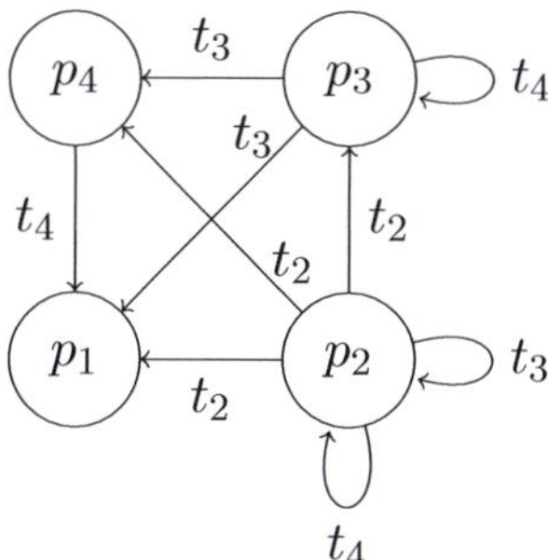


We notice that the transition $t_4$ applied on $p_4$ always leads to $p_1$. We can also see that $p_2$ and $p_3$ are stable by $t_4$. Note that the diagram does not include some adverse transitions, for example $t_2$ applied to $p_4$ sends back to $p_2$. But we can find a winning strategy with the information available in the diagram

- We ask if the starting position is a winning one. If not we are not in $p_1$
- We start by applying $t_4$. If we land in a winning position then we were in $p_4$. If not, we are either in $p_2$ or $p_3$.
- We apply now $t_3$ and then $t_4$. If we land in a winning position (after applying $t_3$ or after applying $t_4$) we can confirm we were in $p_3$ otherwise it means we were in $p_2$
- We know now that we are in $p_2$. We apply $t_2$, $t_3$ and $t_4$ to win.

The winning algorithm including the complementary check is therefore

$$t_4, t_1, t_3, t_1, t_4, t_1, t_2, t_1, t_3, t_1, t_4, t_1$$

## 2.16 N Coins, 1 Table - Solution

**Question :** 2 players take turns placing coins on a large perfectly round table. Coins can not overlap and all the coin surface must be in contact with the table. The first player who can't place a coin loses. Is it better to play first and is there a winning strategy?

**Solution :** The table is round and the winning strategy in this game is based on the central symmetry of the table. The first player A places his coin at the exact center of the table. Every time B places a coin A can respond by placing his coin in the symmetric position. With this strategy B is forced to discover new areas and A is guaranteed to place a coin. Therefore B will eventually run out of space and A is certain to win.

## 2.17 Regression Mirror - Solution

**Question :** Suppose that $X$ and $Y$ are mean zero, unit variance random variables. If least squares regression (without intercept) of $Y$ against $X$ gives a slope of $\beta$ (i.e. it minimises $\mathbb{E}[(Y-\beta X)^2]$), what is the slope of the regression of X against Y?

**Solution :** We know that the slope of the least squares regression of $Y$ against $X$ is

$$\beta = (X^T X)^{-1} X^T Y$$

In this case $X$ and $Y$ are mean zero, unit variance random variables. Therefore

$$X^T X = \text{Var}(X) = \sigma_X^2 = 1$$

$$X^T Y = \text{Cov}(X, Y)$$

$$\beta = \text{Cov}(X, Y)$$

and

$$\beta = \gamma$$

where $\gamma$ is the slope of the regression of $X$ against $Y$.

## 2.18 Bayesian Kids - Solution

**Question :** I meet someone with 2 children, and I learn that one of the children is a boy. What's the probability that the other child is also a boy? What if one of the children is a boy born on a Tuesday?

**Solution :** This is a classic question where the intuitive answer $\left(\frac{1}{2}\right)$ is wrong. To get it right, we write all the possible configurations, they all have a probability equal to $\frac{1}{4}$

$$BB, GG, BG, GB$$

The given information restrains the universe to $BG, GB, BB$ and it appears that the probability that the other child is a boy is actually $\frac{1}{3}$.

Although the provided information seemed symmetric, it actually separated the universe into two blocks where the desired property is skewed. Hence the paradox. Let us see the second case, one of the boys is born on a Tuesday. We create a new property, $X_T$ is a child born a Tuesday and $X_O$ is a child born on another day of the week. Clearly for any child the probability to be born on a Tuesday is $\frac{1}{7}$ and the probability to have a boy born on a Tuesday is $\frac{1}{14}$. The configurations which are compatible with the provided information are now

$$G_T B_T, G_O B_T, B_T G_T, B_T G_O, B_T B_T, B_T B_O, B_O B_T$$

These cases correspond to the event $A$ ={At least one child is a boy born on a Tuesday} and we denote $A^c$ ={No child is a boy born on a Tuesday}. We have

$$P(A^c) = \left(\frac{13}{14}\right)^2$$

$$P(A) = 1 - P(A^c) = \frac{27}{196}$$

and the cases where the other child is also a boy are

$$B_T B_T, B_T B_O, B_O B_T$$

Let us call $C$ this subset of $A$.

$$P(C) = \frac{1}{196} + \frac{6}{196} + \frac{6}{196} = \frac{13}{196}$$

and the desired probability is

$$\frac{P(C)}{P(A)} = \frac{13}{27}$$

## 2.19 Last Digit - Solution

**Question :** Consider all 100 digit numbers, i.e. those between 0 to $(10^{100} - 1)$, inclusive. For each number, take the product of non-zero digits (treat the product of digits of 0 as 1), and sum across all the numbers. What's the last digit?

**Solution :** We consider the problem between 0 and $(10^n - 1)$, let $I_n$ be the sum of the product of the digits in the case $n$. For $n = 1$

$$I_1 = 1 + \sum_{i=1}^{9} i = 46$$

There is an induction relationship between $I_n$ and $I_{n-1}$, when we fix a digit in the case $n + 1$ we are left with the sum of the product of the digits in $n - 1$

$$I_n = 1.I_{n-1} + \sum_{i=1}^{9} i.I_{n-1}$$

Therefore

$$I_n = 46 I_{n-1}$$

and

$$I_n = 46^n$$

To conclude, we notice that if two integers end with a 6, their product ends with a 6. The last digit is therefore a 6.

## 2.20 Repeated Contraction - Solution

**Question :** Let $R(n)$ be a random draw of integers between 0 and $n-1$ (inclusive). I repeatedly apply $R$, starting at $10^{100}$. What's the expected number of repeated applications until I get zero?

**Solution :** Let $F_n$ be the expected number of repeated applications to reach zero when we start at $n$.

$$F_n = 1 + \frac{1}{n}(F_{n-1} + F_{n-2} + \cdots + F_0)$$

where $F_0 = 0$. And

$$F_{n-1} = 1 + \frac{1}{n-1}(F_{n-2} + F_{n-3} + \cdots + F_0)$$

Therefore

$$nF_n - (n-1)F_{n-1} = 1 + F_{n-1}$$

$$F_n = \frac{1}{n} + F_{n-1}$$

So $F_0 = 0$ and when $n > 0$

$$F_n = \sum_{i=1}^{n} \frac{1}{i} \simeq \ln(n)$$

and for $N = 10^{100}$

$$F_N \simeq 100\ln(10) \simeq 230$$

## 2.21 Domino's Pizza - Solution

**Question :** How many ways are there to tile dominos (with size 2 $\times$ 1) on a grid of 2 $\times$ n? How about on a grid of 3 $\times$ 2n?

**Solution :** It is clear that in the case 2 x 1 the answer is 1, and in the case 2 x 2 the answer is 2.

The case 2 x n can be separated into 2 sub-cases. The tiling can either end with a horizontal or a vertical domino

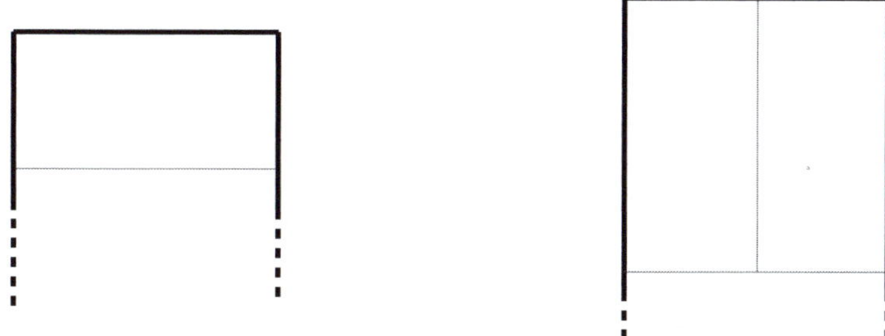

We denote $I_n$ the total number of tiling for 2 x n. $H_n$ and $V_n$ the number of tiling ending respectively with a horizontal and a vertical domino. We have the following induction formula

$$I_n = H_n + V_n = I_{n-1} + I_{n-2}$$

We find that $I_n = \text{Fib}_{n+1}$ is a shifted Fibonacci sequence. The sequence equation is

$$r^2 - r - 1 = 0$$

with roots

$$r_1 = \frac{1+\sqrt{5}}{2}, \; r_2 = \frac{1-\sqrt{5}}{2}$$

the Fibonacci sequence verifies

$$\text{Fib}_n = \frac{1}{\sqrt{5}} r_1^n - \frac{1}{\sqrt{5}} r_2^n$$

and therefore

$$I_n = \frac{1}{\sqrt{5}} r_1^{n+1} - \frac{1}{\sqrt{5}} r_2^{n+1}$$

To solve the configuration 3 x 2n, we start with 3 x 2. There are 3 possible tiling options

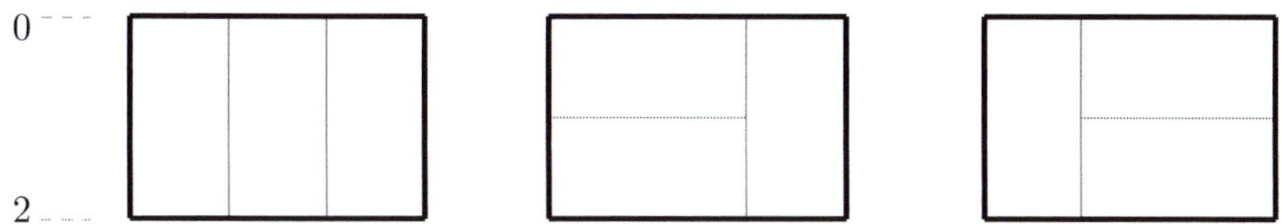


In fact, in the general case 3 x 2n, the tiling has to end with one of these 3 configurations

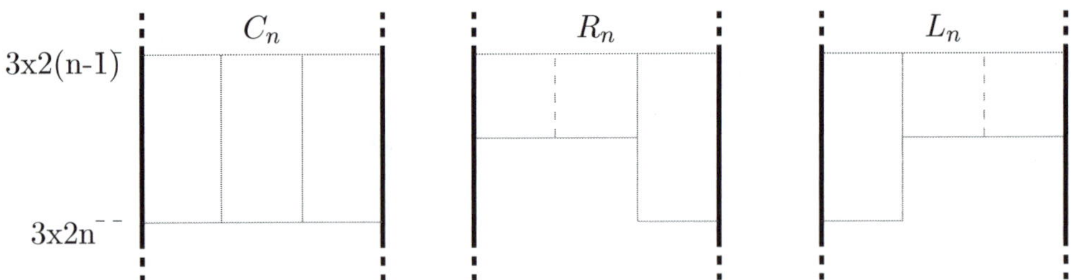


We denote $C_n$, $L_n$ and $R_n$ the number of configurations ending respectively with 3 vertical dominos (Center), only 1 vertical domino on the left and only 1 vertical domino on the right. We call an admissible configuration a full tiling of the surface without holes or overlapping dominos. Note that with the above definition, an $R_n$ configuration can either lead to an admissible (closed) 3x2n configuration or to an overlapping 3x2n configuration as seen in the figure below. The same is true for $L_n$.

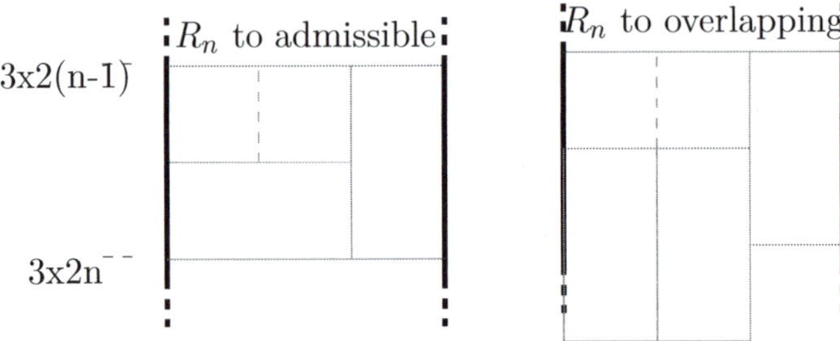


We denote $J_n$ the total number of admissible configurations in the case 3 x 2n. As seen above, every $R_n$ and every $L_n$ configuration can each lead to one and only one admissible configuration. Also $C_n$ is an admissible configuration. So we have $J_n$

$$J_n = C_n + L_n + R_n$$

By symmetry $L_n = R_n$ and

$$J_n = C_n + 2R_n$$

Clearly, every 3 x $2(n-1)$ admissible configuration can be extended with a center pattern to produce an admissible 3 x 2n pattern. It is the only way to create a center pattern. Therefore

$$C_n = J_{n-1}$$

However, we can construct a non centered (Left or Right) configuration with 2 methods: from an admissible configuration or from an overlapping configuration. The figure below illustrates the Right case

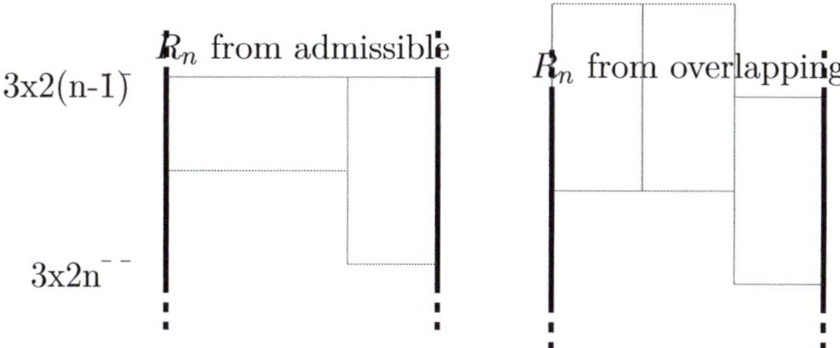


Therefore there is one unique way to construct a $R_n$ configuration from an admissible 3 x $2(n-1)$ configuration, and one unique way from a $R_{n-1}$ configuration. Therefore

$$R_n = J_{n-1} + R_{n-1}$$

and the system of equations

$$\begin{cases} J_n = J_{n-1} + 2R_n \\ R_n = J_{n-1} + R_{n-1} \end{cases}$$

We substitute the second equation in the first equation

$$J_n = J_{n-1} + 2J_{n-2} + 2R_{n-2}$$

and by repeating

$$J_n = J_{n-1} + 2J_{n-2} + 2J_{n-3} + \cdots + 2J_1 + 2R_1$$

which is also true if we start from $n-1$

$$J_{n-1} = J_{n-2} + 2J_{n-3} + 2J_{n-4} + \cdots + 2J_1 + 2R_1$$

by subtracting the two equations

$$J_n - J_{n-1} = J_{n-1} + J_{n-2}$$

$$J_n = 2J_{n-1} + J_{n-2}$$

This is a recursive sequence of order 2 with equation

$$r^2 - 2r - 1 = 0$$

$$r_1 = 1 + \sqrt{2},\ r_2 = 1 - \sqrt{2}$$

We solve

$$J_n = Ar_1^n + Br_2^n,\ J_1 = 3,\ J_2 = 11$$

and we find

$$J_n = \left(\frac{5-\sqrt{2}}{2}\right)\left(1+\sqrt{2}\right)^n + \left(\frac{5+\sqrt{2}}{2}\right)\left(1-\sqrt{2}\right)^n$$

## 2.22 Nash's Car - Solution

**Question :** A company has a competition to win a car. Each contestant needs to pick a positive integer. If there's at least one unique choice, the person who made the smallest unique choice wins the car. If there are no unique choices, the company keeps the car and there's no repeat of the competition. It turns out that there are only three contestants, and you're one of them. Everyone knows before picking their numbers that there are only three contestants. How should you make your choice?

**Solution :** This question is a variation of the Game Theory question (see 1.8). A tempting strategy for all players is to choose the smallest number. But by doing so, they increase their chance to choose the same number as a competitor. The key in this type of question is to observe that all players have access to the same amount of information. Therefore all players will guess the optimal strategy and take full advantage of it. We denote $p(i)$ the discrete probability distribution for every player to pick the number $i$ for his choice. He wins the car if both competitors chose a larger number, or if both competitors chose the same smaller number. His expected gain is

$$\mathbb{E}\left(\text{winnings} \mid \text{player picks } i\right) = E(i) = X\left(\sum_{j=1}^{i-1} p(j)^2 + \left(\sum_{j=i+1}^{\infty} p(j)\right)^2\right) \tag{3}$$

where $X$ is the price of the car. The optimal distribution should make the expected winnings independent of the chosen number. We search for a solution of the form

$$p(i) = Aq^i$$

where $q < 1$ and

$$A = \frac{1}{\sum_{i=1}^{\infty} q^i} = \frac{1-q}{q}$$

We inject it in equation (3)

$$E(i) = XA^2\left(\sum_{j=1}^{i-1} q^{2j} + \left(\sum_{j=i+1}^{\infty} q^j\right)^2\right)$$

$$E(i) = XA^2\left(q^2\frac{1-q^{2i-2}}{1-q^2} + \left(q^{i+1}\frac{1}{1-q}\right)^2\right)$$

$$E(i) = XA^2\left(\frac{(q^2-q^{2i})(1-q)+q^{2i+2}(1+q)}{(1-q)^2(1+q)}\right)$$

$$E(i) = XA^2 \left( \frac{1-q}{1+q} + q^{2i+1} \frac{q^3+q^2+q-1}{1+q} \right)$$

This quantity is constant when $q$ is the only real root of

$$(q^3 + q^2 + q - 1) = 0$$

$$q_0 \simeq 0.5437$$

And we find that the optimal strategy is to pick the positive integer $i$ with a probability

$$p(i) = (1 - q_0) q_0^{i-1}$$

## 2.23 Correlation Impossible I - Solution

**Question :** If X, Y and Z are three random variables such that X and Y have a correlation of 0.9, and Y and Z have correlation of 0.8, what are the minimum and maximum correlation that X and Z can have?

**Solution :** This is a classic question, we want to find the values of Corr$(X, Z)$ for which the following matrix is a valid correlation matrix

$$\begin{pmatrix} 1 & 0.9 & x \\ 0.9 & 1 & 0.8 \\ x & 0.8 & 1 \end{pmatrix}$$

It is a valid correlation matrix if and only if it is positive semi definite. This is equivalent to having no negative eigenvalue. We derive the matrix characteristic polynomial

$$P(\alpha) = \begin{bmatrix} 1-\alpha & 0.9 & x \\ 0.9 & 1-\alpha & 0.8 \\ x & 0.8 & 1-\alpha \end{bmatrix}$$

$$P(\alpha) = (1-\alpha)^3 + 1.44x - (1-\alpha)(1.45 + x^2)$$

The roots of $P$ verify

$$(1-\alpha)^3 = (1-\alpha)(1.45 + x^2) - 1.44x$$

and $\alpha \geq 0$ is equivalent to $(1-\alpha)^3 \leq 1$. Therefore our condition is

$$(1-\alpha)(1.45 + x^2) - 1.44x \leq 1$$

$$0.45 + x^2 - 1.44x \leq 1.45\alpha$$

and we can use again $\alpha \geq 0$

$$Q(x) = x^2 - 1.44x + 0.45 \leq 0$$

The roots of $Q$ are $a \simeq 0.458$ and $b \simeq 0.982$, and the acceptable interval for the correlation between X and Z is

$$I = [a, b]$$

We can also solve this problem with an algebraic approach. For simplicity we can assume that var(X) = var(Y) = var(Z) = 1, as we could normalize the vectors.

$$\text{cov}(X, Y) = \frac{\text{cov}(X, Y)}{\sigma_X \sigma_Y} = \rho_{XY} = 0.9$$

$$\text{cov}(Y, Z) = \frac{\text{cov}(Y, Z)}{\sigma_Y \sigma_Z} = \rho_{YZ} = 0.8$$

$X$ can be projected on $Y$ and its orthogonal space

$$X = \rho_{XY} Y + \sqrt{1 - \rho_{XY}^2} U$$

$Z$ can be projected on $Y$ and its orthogonal space

$$Z = \rho_{YZ} Y + \sqrt{1 - \rho_{YZ}^2} W$$

where $\text{cov}(U, Y) = \text{cov}(W, Y) = 0$ and $\text{var}(U) = \text{var}(W) = 1.$ so we have

$$\text{cov}(Z, X) = \rho_{XY}\rho_{YZ} + \sqrt{(1 - \rho_{XY}^2)(1 - \rho_{YZ}^2)}\text{cov}(U, W)$$

finally we have

$$-1 \leq \text{cov}(U, W) \leq 1$$

leading to

$$0.72 - 0.262 = 0.458 \leq x \leq 0.982 = 0.72 + 0.262$$

## 2.24 Correlation Impossible II - Solution

**Question :** If $X_1, X_2 ... X_n$ are n random variables such that

$$\text{Corr}(X_i, X_j) = \rho \text{ for } i \neq j$$

what are the minimum and maximum values that $\rho$ can have?

**Solution :** We consider the matrix $A$

$$A = \begin{pmatrix} 1 & \rho & \dots & \rho \\ \rho & 1 & \ddots & \vdots \\ \vdots & \ddots & \ddots & \rho \\ \rho & \dots & \rho & 1 \end{pmatrix}$$

It is a valid correlation matrix if and only if it is positive semidefinite. This is equivalent to having no negative eigenvalue. For large matrices with patterns the eigenvalues can often be found visually. We notice that

$$A\begin{pmatrix}1\\1\\\vdots\\1\end{pmatrix}=AX=\begin{pmatrix}1+(n-1)\rho\\1+(n-1)\rho\\\vdots\\1+(n-1)\rho\end{pmatrix}=(1+(n-1)\rho)\,X$$

We notice another pattern when we subtract two lines of $A$. Let $Y_i, i>1$ the family of vectors defined as

$$\begin{cases}Y_i[1]=1\\Y_i[i]=-1\\Y_i[j]=0 \text{ otherwise}\end{cases}$$

We have

$$AY_i=(1-\rho)Y_i$$

The family of vectors $(X, Y_2\dots Y_n)$ is therefore an independent family of $n$ eigenvectors and the eigenvalues of $A$ are $((1-\rho), 1+(n-1)\rho)$. Hence the valid values of $\rho$ are

$$1+(n-1)\rho\geq 0$$

$$\rho\geq -\frac{1}{n-1}$$

## 2.25 The Dark Side of the Die - Solution

**Question :** How many times do I have to roll a die until all six sides appear?

**Solution :** The trick for this question is to consider the expected number of rolls to see a new side. We denote respectively $P_i$ ($E_i$) the probability (the expected number of rolls) to see a new side when we have already seen $i$ sides. It is relatively straightforward that

$$E_i=\frac{1}{P_i}$$

When we start, the probability to see a new side at the next roll is 1. So $P_0=E_0=1$. When we have seen $i$ sides, we have $P_i=\frac{6-i}{6}$. Therefore the total expected number of rolls to see all sides $E$ is

$$E=\sum_0^5 E_i=\sum_0^5\frac{1}{P_i}=\sum_0^5\frac{6}{6-i}=\frac{6}{6}+\frac{6}{5}\dots+\frac{6}{1}=14.7$$

P.S. We denote $E$ the expected number of attempts to make an event of probability $p$ happen. We have

$$E = \sum_{i=1}^{+\infty} ip(1-p)^{i-1}$$

We identify the Taylor development of $\frac{1}{(1-x)^2}$ (see page 128) and

$$E = \frac{p}{(1-(1-p))^2} = \frac{1}{p}$$

## 2.26 Bonus Day - Solution

**Question :** Five pirates $P_i$ have 100 gold coins. They have to divide up the loot. In order of seniority (suppose pirate $P_5$ is most senior, $P_1$ is least senior), the most senior pirate proposes a distribution of the loot. They vote and if at least 50% accept the proposal, the loot is divided as proposed. Otherwise the most senior pirate is executed, and they start over again with the next senior pirate. Which solution does the most senior pirate propose? Assume they are very intelligent and extremely greedy (and that they would prefer not to die).

**Solution :** As it is often the case in this type of questions, we need to work it out backwards starting with smaller systems.

- If there was only 1 pirate, he would take 100 coins.
- In a system with 2 pirates, the most senior would also take 100 coins as he is guaranteed to win the vote.
- In a system with 3 pirates, $P_3$ cannot have 100 coins because he cannot win the vote alone. But he could take 99 coins if he gives 1 coin to $P_1$. $P_1$ would be making more than what he could expect from smaller systems and would vote for the proposal.
- In a system with 4 pirates, the most senior pirate must convince one pirate to accept the proposal. We know that $P_2$ gets nothing in a 3 pirates system. Therefore $P_4$ can take 99 coins, give 1 coin to $P_2$ and win the vote. With one coin, $P_2$ has the opportunity to lock more than his expected gain in a 3 pirates system and he would accept the proposal.
- In a system with 5 pirates, $P_5$ needs to convince 2 pirates to vote with him. He will pick the pirates who would get nothing in a 4 pirates system: $P_3$ and $P_1$.

The most senior pirate should take 98 coins, give 1 coin to $P_3$ and 1 coin to $P_1$.

## 2.27 Secret Polynomial - Solution

**Question :** We consider a polynomial $P(x)$ of which all coefficients are positive integers ($a_i \geq 0$). The polynomial is in a black box and we can only retrieve its value in given points. In how many points do we need to value the polynomial in order to find the values of all the coefficients?

**Solution :** The answer is 2. Let $P(x) = \sum_{i=0}^{n} a_i x^i$ be the polynomial, we start by taking the value of $P(1)$. $P(1)$ is greater than any coefficient $a_i$ because all the coefficients are positive. We denote $M = P(1) + k$ where $k > 0$. Now we take the value of $P(M) = \sum_{i=0}^{n} a_i M^i$. The coefficients $a_i$ are therefore the coefficients of the decomposition of $P(M)$ in base $M$ and can now be found with a sequence of euclidean divisions.

## 2.28 Drunk Mutant Ninja Ant - Solution

**Question :** An ant starts a walk from a cube vertex, it walks on the edges and at every vertex it chooses to walk one of the available edges (including the edge it came from) with an equal probability. How many edges will the ant cross in average to come back to the starting point?

**Solution :** We assign numbers to the cube vertices and we denote 1 the vertex from which the ant starts its journey

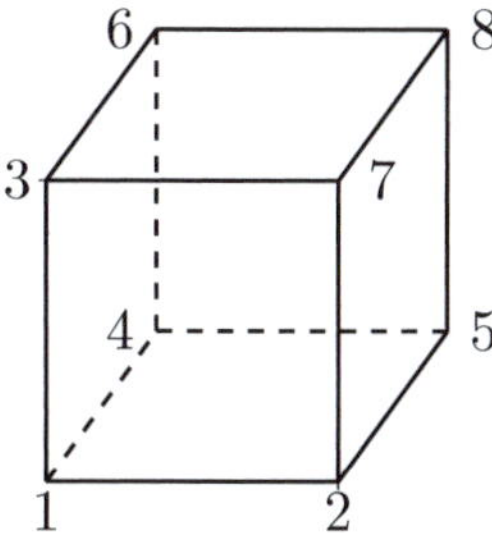


We can divide the vertices into groups or layers with interesting properties. We denote $L_1 = \{1\}$, $L_2 = \{2, 3, 4\}$, $L_3 = \{5, 6, 7\}$ and $L_4 = \{8\}$. We notice that the ant always has to change layer when it travels through an edge. We can draw the following transition diagram.

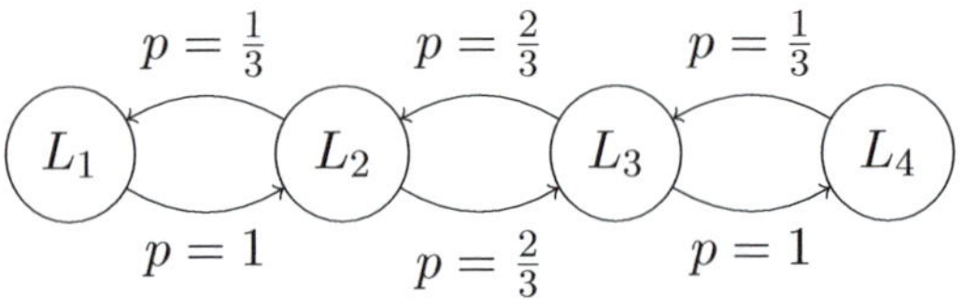

We denote now $N_i$ the average number of edges the ant needs to travel to come back to the vertex 1 given that it starts from the layer $i$. We have the system

$$\begin{cases} N_1 = 1 + N_2 \\ N_2 = \frac{1}{3} + \frac{2}{3}(1 + N_3) \\ N_3 = \frac{2}{3}(1 + N_2) + \frac{1}{3}(1 + N_4) \\ N_4 = 1 + N_3 \end{cases}$$

We can eliminate $N_1$ and $N_4$

$$\begin{cases} N_2 = \frac{1}{3} + \frac{2}{3}(1 + N_3) \\ N_3 = \frac{2}{3}(1 + N_2) + \frac{1}{3}(2 + N_3) \end{cases}$$

we solve to find

$$\begin{cases} N_1 = 8 \\ N_2 = 7 \\ N_3 = 9 \\ N_4 = 10 \end{cases}$$

And the answer is 8.

## 2.29 Dog Day Afternoon - Solution

**Question :** You are standing at the centre of a circular field of radius R. The field has a low wire fence around it. Attached to the wire fence (and restricted to running around the perimeter) is a large, sharp-fanged, hungry dog. You can run at speed $v$, while the dog can run four times as fast. What is your running strategy to escape the field?

**Solution :** This is a surprisingly popular question in interviews. The key is to keep things simple. If you run to the fence from the center and away from the dog, it would leave you running a distance $R$ and the dog running $\pi R$ this would take you a time $R/v$ and him $\pi R/(4v)$ so clearly he would get there before you as

$$\frac{R}{v} > \frac{\pi R}{4v}$$

First let us find the smallest distance from the center $R_1$ from which running away from the dog is a winning strategy. The dog needs to run $\pi R$ to catch up and $R_1$ verifies

$$\frac{R - R_1}{v} = \frac{\pi R}{4v}$$

and

$$R_1 = R - \frac{\pi R}{4} \simeq 0.215R$$

Now when we are on a smaller circle we can have a higher angular speed than the dog. The other interesting radius is the longest distance from the center $R_2$ for which we have a higher angular speed than the dog. We compare the times to run a full circle to find $R_2$, $R_2$ verifies

$$\frac{2\pi R}{4v} = \frac{2\pi R_2}{v}$$

and

$$R_2 = \frac{R}{4} = 0.25R$$

So the strategy is clear: walk to a distance $R_3$ from the center such that $R_3 \in ]R_1, R_2[$. Then run along the circle of radius $R_3$ until the dog is diametrically opposed to you, it is possible because $R_3 < R_2$. Then run towards the fence, you will make it before the dog because $R_3 > R_1$.

# Chapter 3

# Stochastic Calculus

# Stochastic Calculus

## 3.1 Lognormal Expectation ♠ (Natixis)

Calculate $\mathbb{E}(\exp(X))$ when is $X$ is a normally distributed random variable

$$X \sim \mathcal{N}\left(\mu, \sigma^2\right)$$

---

Solution in page 49

## 3.2 Cumulative Brownian ♠ (Deutsche Bank)

Calculate $\mathbb{E}(\Phi(W_t))$ where $W_t$ a brownian motion and $\Phi$ the standard normal cumulative distribution.

---

Solution in page 50

## 3.3 Multiplicative Itô ♠ (BNP)

For each of the processes $X_t$ below find the process $a(s, \omega)$ such that

$$X_t = E[X_t] + \int_0^t a dB_s$$

i) $X_t = B_t^2$

ii) $X_t = B_t^3$

iii) $X_t = e^{B_t}$

iv) $X_t = \sin B_t$

---

Solution in page 50

## 3.4 Two Sided Corridor ♠♠ (BNP)

Let $B_t$ be a Brownian Motion and $u$ and $d$ two positive real numbers. We consider an option which pays 1 if $B_t$ reaches $u$ and remained greater then $-d$ since inception

$$\exists t_0 : B_{t_0} = u; \ \forall t \in [0, t_0], B_t > -d$$

payment is made when the barrier is touched. Calculate the price of this option when rates are zero.

---

Solution in page 52

## 3.5 One sided corridor ♠♠ (BNP)

Let $B_t$ be a Brownian Motion and $u$ a positive real number. We consider an option which pays 1 if $B_t$ reaches $u$

$$\exists t_0 : B_{t_0} = u$$

payment is made when the barrier is touched. Calculate the price of this option when rates are zero and with rates $r > 0$.

---

Solution in page 53

## 3.6 Two Sided Corridor With Rates ♠♠ (BNP)

Let $B_t$ be a Brownian Motion and $u$ and $d$ two positive real numbers. We consider an option which pays 1 if $B_t$ reaches $u$ and remained greater then $-d$ since inception

$$\exists t_0 : B_{t_0} = u; \ \forall t \in [0, t_0], B_t > -d$$

payment is made when the barrier is touched. Calculate the price of this option with rates $r > 0$.

---

Solution in page 54

## 3.7 Bouncing Corridor ♠♠♠ (BNP)

Let $B_t$ be a Brownian Motion and $u$ and $d$ two positive real numbers. We consider an option which pays 1 if $B_t$ reaches $u$ and touched the down barrier before. The option is knocked out and pays zero if it touches the up barrier first.

$$\text{pays 1 if } \exists t_0 : B_{t_0} = u; \ \exists t_1 \in [0, t_0] : B_{t_1} = -d$$

$$\text{pays 0 if } \exists t_0 : B_{t_0} = u; \ \forall t \in [0, t_0], B_t > -d$$

payment is made when the barrier is touched. Calculate the price of this option with rates $r > 0$. Generalize to an option paying 1 after $n$ bounces (pays 1 if the option touches $u$ then $u$ then $-d$ etc... n times, pays 0 if $u$ is touched first).

---

Solution in page 55

## 3.8 Brownian Bridge ♠ (JP Morgan)

Let $B_s$ be a Brownian bridge, that is a Brownian motion constrained such that $B_0 = 0$ and $B_t = x$. What is the distribution of $B_s$, for $0 \leq s < t$?

---

Solution in page 56

## 3.9 Martingales? ♠ (Societe Generale)

Are the following processes martingales?

i) $X_t = B_t + 8t$

ii) $X_t = B_t^2$

iii) $X_t = B_t^3$

iv) $X_t = t^2 B_t - 2\int_0^t sB_s ds$

Solution in page 57

## 3.10 Natural Martingale ♠♠♠ (JP Morgan)

Let $Y$ be a real valued random variable on $(\Omega, \mathcal{F}, P)$ such that

$$E[|Y|] < \infty$$

Define

$$M_t = E[Y|\mathcal{F}_t]; \quad t \geq 0$$

Show that $M_t$ is an $\mathcal{F}_t$-martingale. Conversely, let $M_t; t \geq 0$ be a real valued $\mathcal{F}_t$-martingale such that

$$\sup_{t \geq 0} E\left[|M_t|^p\right] < \infty \quad \text{for some} \quad p > 1$$

Show that there exists $Y \in L^1(P)$ such that $M_t = E[Y|\mathcal{F}_t]$.

Solution in page 58

## 3.11 Exponential Brownian ♠♠ (JP Morgan)

Let $B_t$ be Brownian motion on $\mathbf{R}, B_0 = 0$. Prove that

$$\mathbb{E}\left[e^{iuB_t}\right] = \exp\left(-\frac{1}{2}u^2 t\right) \qquad \text{for all} \quad u \in \mathbf{R}$$

Use the power series expansion of the exponential function on both sides, compare the terms with the same power of $u$ and deduce that

$$\mathbb{E}\left[B_t^4\right] = 3t^2$$

and more generally that

$$\mathbb{E}\left[B_t^{2k}\right] = \frac{(2k)!}{2^k \cdot k!} t^k; \quad k \in \mathbf{N}$$

Solution in page 59

Chapter 4

# Stochastic Calculus - Solutions

# Stochastic Calculus - Solutions

## 4.1 Lognormal Expectation - Solution

**Question :** Calculate $\mathbb{E}(\exp(X))$ when is $X$ is a normally distributed random variable

$$X \sim \mathcal{N}\left(\mu, \sigma^2\right)$$

**Solution :** The result can be found elegantly, we know that $Y_t$ is a martingale where

$$Y_t = \exp\left(-\frac{\sigma^2}{2}t + \sigma B_t\right) = \exp\left(-\frac{\sigma^2}{2}t + \sigma\sqrt{t}Z\right)$$

where $Z$ is a standard normal distribution. $X' = \mu + \sigma Z$ has the same distribution as $X$, we set $t = 1$

$$\mathbb{E}(Y_1) = \mathbb{E}(Y_0) = 1 = \exp\left(-\frac{\sigma^2}{2} - \mu\right)\mathbb{E}\left(\exp\left(\mu + \sigma Z\right)\right)$$

and therefore

$$\mathbb{E}\left(\exp(X)\right) = \exp\left(\frac{\sigma^2}{2} + \mu\right)$$

The expectation can also be calculated using integrals

$$\mathbb{E}\left(\exp(X)\right) = \int_{-\infty}^{\infty} \exp(x)\frac{1}{\sqrt{2\pi}\sigma}\exp\left(-\frac{(x-\mu)^2}{2\sigma^2}\right)dx$$

$$\mathbb{E}\left(\exp(X)\right) = \frac{1}{\sqrt{2\pi}\sigma}\int_{-\infty}^{\infty} \exp\left(\frac{2\sigma^2 x - x^2 + 2\mu x - \mu^2}{2\sigma^2}\right)dx$$

$$\mathbb{E}\left(\exp(X)\right) = \frac{1}{\sqrt{2\pi}\sigma}\int_{-\infty}^{\infty} \exp\left(\frac{-\left(x - (\sigma^2 + \mu)\right)^2}{2\sigma^2} + \frac{\sigma^4 + 2\mu\sigma^2}{2\sigma^2}\right)dx$$

$$\mathbb{E}\left(\exp(X)\right) = \exp\left(\frac{\sigma^2}{2} + \mu\right)\int_{-\infty}^{\infty} \frac{1}{\sqrt{2\pi}\sigma}\exp\left(\frac{-\left(x - (\sigma^2 + \mu)\right)^2}{2\sigma^2}\right)dx$$

We identify the integral of the density of a normal distribution, equal to 1 and

$$\mathbb{E}\left(\exp(X)\right) = \exp\left(\frac{\sigma^2}{2} + \mu\right)$$

## 4.2 Cumulative Brownian - Solution

**Question :** Calculate $\mathbb{E}(\Phi(W_t))$ where $W_t$ a brownian motion and $\Phi$ the standard normal cumulative distribution.

**Solution :** The elegant solution for this problem is based on the symmetry of the brownian motion:

$$\mathbb{E}\left(\Phi\left(W_t\right)\right) = \mathbb{E}\left(\Phi\left(-W_t\right)\right) = \mathbb{E}\left(1 - \Phi\left(W_t\right)\right) = 1 - \mathbb{E}\left(\Phi\left(W_t\right)\right)$$

and we get

$$\mathbb{E}\left(\Phi\left(W_t\right)\right) = \frac{1}{2}$$

The result can also be proved using integrals

$$\mathbb{E}\left(\Phi\left(W_t\right)\right) = \int_{-\infty}^{+\infty}\left(\int_{-\infty}^{x} \frac{1}{\sqrt{2\pi}} \exp\left(\frac{-u^2}{2}\right) du\right) \frac{1}{\sqrt{2\pi t}} \exp\left(\frac{-x^2}{2t}\right) dt$$

$$\mathbb{E}\left(\Phi\left(W_t\right)\right) = \int_{-\infty}^{+\infty} \Phi(x)\Phi'(x)dx = \left[\frac{\Phi^2(x)}{2}\right]_{-\infty}^{+\infty} = \frac{1}{2}$$

## 4.3 Multiplicative Itô- Solution

**Question :** For each of the processes $X_t$ below find the process $a(s,\omega)$ such that

$$X_t = E[X_t] + \int_0^t a dB_s$$

i) $X_t = B_t^2$

ii) $X_t = B_t^3$

iii) $X_t = e^{B_t}$

iv) $X_t = \sin B_t$

**Solution :**

i) $X_t = B_t^2$

We apply Itô's formula to find the dynamics of $X_t$

$$dX_t = 2B_t dB_t + dt$$

$$X_t = t + \int_0^t 2B_s dB_s$$

Which corresponds to the required decomposition with

$$\mathbb{E}[X_t] = t; \quad a = 2B_s$$

ii) $X_t = B_t^3$

By applying Itô's formula we get

$$dX_t = 3B_t^2 dB_t + 3B_t dt$$

we need to decompose further the term $3B_t dt$, we consider the process $Y_t = tB_t$

$$dY_t = B_t dt + t dB_t$$

$$Y_t = \int_0^t B_s ds + \int_0^t s dB_s$$

therefore

$$\int_0^t B_s ds = tB_t - \int_0^t s dB_s$$

we can reinject this equation in $X_t$

$$X_t = \int_0^t 3B_s^2 dB_s + 3tB_t - 3\int_0^t s dB_s$$

$$X_t = \int_0^t 3B_s^2 dB_s + 3t\int_0^t dB_s - 3\int_0^t s dB_s$$

giving the desired decomposition with

$$\mathbb{E}[X_t] = 0; \quad a = 3B_s^2 + 3(t-s)$$

iii) $X_t = e^{B_t}$

We apply Itô's formula to find the dynamics of $X_t$

$$dX_t = e^{B_t} dB_t + \frac{1}{2} e^{B_t} dt$$

We can eliminate the term in $dt$ with a classic technique used for Ornsetin Uhlenbeck derivation (see page 101). We denote $A$ a constant and we consider the process $Y_t = e^{At} e^{B_t}$

$$dY_t = e^{B_t} e^{At} dB_t + \frac{1}{2} e^{B_t} e^{At} dt + A e^{B_t} e^{At} dt$$

we choose $A = -\frac{1}{2}$ and we get

$$d\left(e^{B_t} e^{\frac{-t}{2}}\right) = e^{B_t} e^{\frac{-t}{2}} dB_t$$

$$e^{B_t}e^{\frac{-t}{2}} - 1 = \int_0^t e^{B_s}e^{\frac{-s}{2}}dB_s$$

and

$$e^{B_t} = e^{\frac{t}{2}} + \int_0^t e^{B_s+\frac{t-s}{2}}dB_s$$

giving the decomposition

$$\mathbb{E}[X_t] = e^{\frac{t}{2}}; \quad a = e^{B_s+\frac{t-s}{2}}$$

iv) $X_t = \sin B_t$
We start with Itô's formula to find the dynamics of $X_t$

$$dX_t = \cos B_t dB_t - \frac{1}{2}\sin B_t dt$$

We can eliminate the term in $dt$ by introducing $Y_t = e^{\frac{t}{2}}\sin B_t$ (see previous case).

$$d\left(\sin B_t e^{\frac{t}{2}}\right) = \cos B_t e^{\frac{t}{2}}dB_t$$

and

$$\sin B_t = \int_0^t \cos B_s e^{\frac{s-t}{2}}dB_s$$

giving the decomposition

$$\mathbb{E}[X_t] = 0; \quad a = \cos B_s e^{\frac{s-t}{2}}$$

## 4.4 Two Sided Corridor - Solution

**Question :** Let $B_t$ be a Brownian Motion and $u$ and $d$ two positive real numbers. We consider an option which pays 1 if $B_t$ reaches $u$ and remained greater then $-d$ since inception

$$\exists t_0 : B_{t_0} = u; \; \forall t \in [0, t_0], B_t > -d$$

payment is made when the barrier is touched. Calculate the price of this option when rates are zero.

**Solution :** This is a classic application of the optional sampling theorem (see page 123). We define $\tau$ the first hitting time of $u$ or $-d$. $\tau$ is a stopping time. The process $B_{\tau\wedge t}$ is a bounded martingale. By applying the optional sampling theorem to $B_{\tau\wedge t}$ and $\tau$ we obtain

$$\mathbb{E}(B_{\tau\wedge T}) = \mathbb{E}(B_\tau) = \mathbb{E}(B_0) = 0$$

and

$$\mathbb{E}(B_\tau) = p.u - (1-p).d = 0$$

where $p$ is the probability to hit $u$ first. The price of the option is therefore

$$\text{Price} = P\{\text{hit } u \text{ first}\} = \frac{d}{u+d}$$

## 4.5 One Sided Corridor - Solution

**Question :** Let $B_t$ be a Brownian Motion and $u$ a positive real number. We consider an option which pays 1 if $B_t$ reaches $u$

$$\exists t_0 : B_{t_0} = u$$

payment is made when the barrier is touched. Calculate the price of this option when rates are zero and with rates $r > 0$.

**Solution :** In this version of the question the stopped Brownian motion is not bounded and we cannot directly apply the optional sampling theorem. Actually the probability of reaching $u$ is 1 because the probability of the brownian motion reaching any given point is 1 (see page 124). The price of the option without rates is 1. When interest rates are applied we need to evaluate $\mathbb{E}(\exp(-r\tau_u))$ where $\tau_u$ is the first hitting time of $u$. In order to evaluate it we consider the martingale

$$X_t = \exp\left(aB_t - \frac{a^2}{2}t\right)$$

with $a > 0$. The process $X_{\tau\wedge t}$ is a bounded martingale. By applying the optional sampling theorem (see page 123) to $X_{\tau_u\wedge t}$ and $\tau_u$ we obtain

$$\mathbb{E}(X_{\tau_u\wedge\tau_u}) = \mathbb{E}(X_{\tau_u}) = \mathbb{E}(X_0) = 1$$

and

$$\mathbb{E}(X_{\tau_u}) = \mathbb{E}\left(\exp\left(a.u - \frac{a^2}{2}\tau_u\right)\right) = 1$$

$$\mathbb{E}\left(\exp\left(-\frac{a^2}{2}\tau_u\right)\right) = \exp(-a.u)$$

We set $a = \sqrt{2r}$ to find the option price

$$\text{Price} = \mathbb{E}(\exp(-r\tau_u)) = \exp\left(-\sqrt{2r}.u\right)$$

NB. The one-sided version of the corridor can be a confusing question. It is sometimes asked differently with a stock following brownian motion dynamics, starting in $A$ and paying 1 if the stock reaches $B > A$. The interviewer often assumes that the process behaves like a stock in the sense that if it touches zero the process stays at zero (the stock of a company is equal to zero in case of default). That version is equivalent to a double sided corridor between 0 and $B$.

## 4.6 Two Sided Corridor With Rates - Solution

**Question :** Let $B_t$ be a Brownian Motion and $u$ and $d$ two positive real numbers. We consider an option which pays 1 if $B_t$ reaches $u$ and remained greater then $-d$ since inception

$$\exists t_0 : B_{t_0} = u;\ \forall t \in [0, t_0], B_t > -d$$

payment is made when the barrier is touched. Calculate the price of this option with rates $r > 0$.

**Solution :** We define $\tau$ the first hitting time of $u$ or $-d$. Let $U$ be the subset of $\Omega$ where $u$ is hit first and $D$ be the subset where $-d$ is hit first. We need to evaluate $\mathbb{E}\left(\mathbb{1}_U \exp\left(-r\tau\right)\right)$. We consider the martingale

$$X_t = \exp\left(-rt + \sqrt{2r}B_t\right)$$

$X_{\tau \wedge t}$ is a bounded martingale. By applying the optional sampling theorem (see page 123) to $X_{\tau_u \wedge t}$ and $\tau_u$ we obtain

$$\mathbb{E}(X_\tau) = \mathbb{E}\left(\mathbb{1}_U \exp\left(-r\tau + \sqrt{2r}u\right)\right) + \mathbb{E}\left(\mathbb{1}_D \exp\left(-r\tau - \sqrt{2r}d\right)\right) = 1$$

But this is also true for the process

$$Y_t = \exp\left(-rt - \sqrt{2r}B_t\right)$$

yielding the equation system

$$\exp\left(\sqrt{2r}u\right) A_u + \exp\left(-\sqrt{2r}d\right) A_d = 1$$

$$\exp\left(-\sqrt{2r}u\right) A_u + \exp\left(\sqrt{2r}d\right) A_d = 1$$

where $A_u = \mathbb{E}\left(\mathbb{1}_U \exp\left(-r\tau\right)\right)$ and $A_d = \mathbb{E}\left(\mathbb{1}_D \exp\left(-r\tau\right)\right)$. Solving the system we find

$$A_u = \frac{\sinh\left(\sqrt{2r}u\right)}{\sinh\left(\sqrt{2r}(u+d)\right)}$$

$$A_d = \frac{\sinh\left(\sqrt{2r}d\right)}{\sinh\left(\sqrt{2r}(u+d)\right)}$$

and in this case

$$\text{Price} = A_u = \mathbb{E}\left(\mathbb{1}_U \exp\left(-r\tau\right)\right)$$

## 4.7 Bouncing Corridor - Solution

**Question :** Let $B_t$ be a Brownian Motion and $u$ and $d$ two positive real numbers. We consider an option which pays 1 if $B_t$ reaches $u$ and touched the down barrier before. The option is knocked out and pays zero if it touches the up barrier first.

$$\text{pays 1 if } \exists t_0 : B_{t_0} = u;\ \exists t_1 \in [0, t_0] : B_{t_1} = -d$$

$$\text{pays 0 if } \exists t_0 : B_{t_0} = u;\ \forall t \in [0, t_0], B_t > -d$$

payment is made when the barrier is touched. Calculate the price of this option with rates $r > 0$. Generalize to an option paying 1 after $n$ bounces (pays 1 if the option touches $u$ then $u$ then $-d$ etc... n times, pays 0 if $u$ is touched first).

**Solution :** We define $\tau$ the first hitting time of $u$ or $-d$. $\tau$ is a stopping time. If $u$ is touched first the payoff is zero. If $-d$ is touched first the option becomes similar to the one sided barrier case with an upper barrier at $(u+d)$. The price of that option was calculated in 3.5

$$\text{Price}_{\text{one sided}} = \exp\left(-\sqrt{2r}(u+d)\right)$$

and the price of the bouncing option is

$$\text{Price} = \exp\left(-\sqrt{2r}(u+d)\right)\mathbb{E}\left(\mathbb{1}_D \exp\left(-r\tau\right)\right)$$

where $D$ be the subset where $-d$ is hit first. The expectation term was calculated in the question 3.6 and we get

$$\text{Price} = \exp\left(-\sqrt{2r}(u+d)\right)\frac{\sinh\left(\sqrt{2r}d\right)}{\sinh\left(\sqrt{2r}(u+d)\right)}$$

To generalize we add one bounce, we consider an option paying 1 if the process touches consecutively $u$, $-d$ and $u$. In this case when $u$ is touched we have a new type of option. We will receive 1 after the down barrier and the up barrier are touched consecutively but there is no knockout feature. We price this option first. We consider a different Brownian Motion $W_t$ starting at 0 and paying 1 if a down barrier at $(-u-d)$ is touched and $W_t$ returns to 0. We denote $\tilde{\tau}$ the first hitting time of $(-u-d)$.

$$\text{Price}_{\text{no knockout}} = \exp\left(-\sqrt{2r}(u+d)\right)\mathbb{E}\left(\exp\left(-r\tilde{\tau}\right)\right)$$

$$\text{Price}_{\text{no knockout}} = \exp\left(-2\sqrt{2r}(u+d)\right)$$

and the price of the option with 2 bounces is

$$\text{Price}_{\text{2 bounces}} = \exp\left(-2\sqrt{2r}(u+d)\right)\mathbb{E}\left(\mathbb{1}_U \exp\left(-r\tau\right)\right)$$

$$\text{Price}_{\text{2 bounces}} = \exp\left(-2\sqrt{2r}(u+d)\right)\frac{\sinh\left(\sqrt{2r}u\right)}{\sinh\left(\sqrt{2r}(u+d)\right)}$$

and generalizing to $n$ bounces, if the first triggering barrier is up

$$\text{Price}_{\text{n bounces}} = \exp\left(-n\sqrt{2r}(u+d)\right)\frac{\sinh\left(\sqrt{2r}u\right)}{\sinh\left(\sqrt{2r}(u+d)\right)}$$

and when the first triggering barrier is down

$$\text{Price}_{\text{n bounces}} = \exp\left(-n\sqrt{2r}(u+d)\right)\frac{\sinh\left(\sqrt{2r}d\right)}{\sinh\left(\sqrt{2r}(u+d)\right)}$$

## 4.8 Brownian Bridge - Solution

**Question :** Let $B_s$ be a Brownian bridge, that is a Brownian motion constrained such that $B_0 = 0$ and $B_t = x$. What is the distribution of $B_s$, for $0 \leq s < t$?

**Solution :** We are looking for the distribution of $B_s$. We denote $\Delta y = [y, y+dy]$ and we consider the infinitesimal probability

$$\mathbb{P}\left(B_s \in \Delta y | B_t \in \Delta x \text{ and } B_0 = 0\right)$$

$B_0$ is deterministic, we apply Bayes Formula (see page 125)

$$\mathbb{P}\left(B_s \in \Delta y | B_t \in \Delta x\right) = A = \frac{\mathbb{P}\left(B_s \in \Delta y, B_t \in \Delta x\right)}{\mathbb{P}\left(B_t \in \Delta x\right)}$$

The increments of the Brownian Motion being independent we have

$$A = \frac{\mathbb{P}\left(B_s \in \Delta y\right)\mathbb{P}\left(\left(B_t - B_s\right) \in \Delta(y-x)\right)}{\mathbb{P}\left(B_t \in \Delta x\right)}$$

$$\mathbb{P}\left((B_t - B_s) \in \Delta(y-x)\right) = \frac{1}{\sqrt{2\pi(t-s)}} \exp\left(-\frac{(x-y)^2}{2(t-s)}\right) dx$$

$$A = \frac{1}{\sqrt{2\pi s}} \exp\left(-\frac{y^2}{2s}\right) \frac{1}{\sqrt{2\pi(t-s)}} \exp\left(-\frac{(x-y)^2}{2(t-s)}\right) \sqrt{2\pi t} \exp\left(\frac{x^2}{2t}\right)$$
$$= \frac{\sqrt{t}}{\sqrt{2\pi s(t-s)}} \exp\left(-\frac{(y-\frac{s}{t}x)^2}{\frac{s(t-s)}{t}}\right)$$

We find that $B_s$ is normally distributed with mean $\frac{xs}{t}$ and variance $\frac{s(t-s)}{t}$.

## 4.9 Martingales? - Solution

**Question :** Are the following processes martingales?

i) $X_t = B_t + 8t$

ii) $X_t = B_t^2$

iii) $X_t = B_t^3$

iv) $X_t = t^2 B_t - 2\int_0^t sB_s ds$

**Solution :**

i) $X_t = B_t + 8t$
In this case the expectation is clearly not constant

$$\mathbb{E}(X_t) = \mathbb{E}(B_t) + 8t = 8t$$

Therefore $X_t$ is not a martingale

ii) $X_t = B_t^2$
Again the expectation is not constant

$$\mathbb{E}(B_t^2) = \text{Var}(B_t) - \mathbb{E}(B_t)^2 = t$$

Therefore $X_t$ is not a martingale

iii) $X_t = B_t^3$
In this case the expectation is zero (the distribution function of $B_t^3$ is even). We can come back to the definition of a martingale, let $t > s$

$$B_t^3 = (B_t - B_s + B_s)^3 = (B_t - B_s)^3 + 3B_s(B_t - B_s)^2 + 3B_s^2(B_t - B_s) + B_s^3$$

we take the conditional expectation and we use

$$\mathbb{E}[(B_t - B_s)^3 | Fs] = \mathbb{E}[(B_t - B_s) | Fs] = 0$$

and we find

$$\mathbb{E}[B_t^3 | Fs] = B_s^3 + 3(t-s)B_s$$

Therefore $X_t$ is not a martingale

iv) $X_t = t^2 B_t - 2\int_0^t sB_s ds$
We apply Itô on $Y_t = t^2 B_t$

$$dY_t = 2tB_t dt + t^2 dB_t$$

and we find

$$X_t = \int_0^t s^2 dB_s$$

Therefore $X_t$ is a martingale

## 4.10 Natural Martingale - Solution

**Question :** Let $Y$ be a real valued random variable on $(\Omega, \mathcal{F}, P)$ such that

$$E[|Y|] < \infty$$

Define

$$M_t = E[Y|\mathcal{F}_t]; \quad t \geq 0$$

Show that $M_t$ is an $\mathcal{F}_t$-martingale. Conversely, let $M_t; t \geq 0$ be a real valued $\mathcal{F}_t$-martingale such that

$$\sup_{t\geq 0} E\left[|M_t|^p\right] < \infty \text{ for some } p > 1$$

Show that there exists $Y \in L^1(P)$ such that $M_t = E[Y|\mathcal{F}_t]$.

**Solution :** We prove that $M_t$ is martingale using the tower rule, let $t > s$

$$\mathbb{E}[M_t|\mathcal{F}s] = \mathbb{E}[\mathbb{E}[Y|\mathcal{F}t]|\mathcal{F}s] = \mathbb{E}[Y|\mathcal{F}s] = M_s$$

To prove the existence of $Y$ we use the Doob's Martingale Convergence Theorem (see page 126). $M_t$ is uniformly integrable with the test function $\psi(x) = x^p$, therefore there exists $Y \in L^1(P)$ such that $M_t \to Y$ a.e. $(P)$ and

$$\int |M_t - Y|\, dP \to 0 \text{ as } t \to \infty$$

therefore for $s > t$

$$\mathbb{E}[Y|\mathcal{F}_t] = \lim_{s\to\infty} \mathbb{E}[M_s|\mathcal{F}_t] = M_t$$

## 4.11 Exponential Brownian - Solution

**Question :** Let $B_t$ be Brownian motion on $\mathbf{R}$, $B_0 = 0$. Prove that

$$\mathbb{E}\left[e^{iuB_t}\right] = \exp\left(-\frac{1}{2}u^2 t\right) \qquad \text{for all } \ u \in \mathbf{R}$$

Use the power series expansion of the exponential function on both sides, compare the terms with the same power of $u$ and deduce that

$$\mathbb{E}\left[B_t^4\right] = 3t^2$$

and more generally that

$$\mathbb{E}\left[B_t^{2k}\right] = \frac{(2k)!}{2^k \cdot k!}t^k; \quad k \in \mathbf{N}$$

**Solution :** We use the geometric brownian motion, which we known is a martingale with dynamics $dX_t = \sigma X_t dB_t$

$$\mathbb{E}(X_t) = \mathbb{E}\left(\exp\left(-\frac{1}{2}\sigma^2 t + \sigma B_t\right)\right) = 1$$

and we set $\sigma = iu$ to obtain

$$\mathbb{E}\left[e^{iuB_t}\right] = \exp\left(-\frac{1}{2}u^2 t\right)$$

we use a Taylor expansion on both sides

$$\sum_{i\in\mathbf{N}} \mathbb{E}\left(\frac{(iuB_t)^n}{n!}\right) = \sum_{i\in\mathbf{N}} \frac{(-u^2t)^n}{2^n n!}$$

The equation above holds for any $u \in \mathbf{R}$, therefore we can identify the terms with the same power of $u$

$$\mathbb{E}\left(\frac{(iuB_t)^{2n}}{(2n)!}\right) = \frac{(-1)^n(u)^{2n}t^n}{2^n n!}$$

$$\mathbb{E}\left(\frac{u^{2n}(-1)^n(B_t)^{2n}}{(2n)!}\right) = \frac{(-1)^n u^{2n} t^n}{2^n n!}$$

and

$$\mathbb{E}\left(B_t^{2n}\right) = \frac{(2n)!}{2^n n!}t^n$$

## Chapter 5

# Finance

# Finance

## 5.1 Binary Hedging ♠♠♠ (UBS)

A trader suggests the following binary hedging strategy for a call option:

- sell a call option at strike $K > S_0$
- buy the stock at $K$ when $S_t$ is increasing and crosses $K$
- sell the stock at $K$ when $S_t$ is decreasing and crosses $K$

What is wrong with this strategy?

---

Solution in page 69

## 5.2 Exchange Option ♠♠ (Credit Suisse)

The payoff of an exchange option at expiry is

$$\text{Ex(T)} = (S_1(T) - S_2(T))^+$$

Calculate the price of an exchange option at $t = 0$ when $\rho$ is the correlation between $S_1$ and $S_2$ and $\sigma$ and $r$ are constant.

---

Solution in page 70

## 5.3 Chooser Option ♠ (Commerzbank)

A chooser option gives the right to choose at some future date $\tau$ to receive a call or put option with strike $K$ and final expiry $T > \tau$. The payoff at $\tau$ of a standard chooser option is

$$\text{Ch}(\tau) = \max\left(C\left(S_\tau, T - \tau, K\right), P\left(S_\tau, T - \tau, K\right)\right)$$

Calculate the price of a chooser option at $t = 0$ when $\sigma$ and $r$ are constant.

---

Solution in page 72

## 5.4 Forward Start Option ♠ (Goldman sachs)

The terminal payoff of a forward start call option is

$$\text{Fs}(T) = (S_T - KS_{T_0})^+$$

where $0 < T_0 < T$. Calculate the price of a forward start call option at $t = 0$ when $\sigma$ and $r$ are constant.

Solution in page 73

## 5.5 Compound Option ♠ (Goldman sachs)

The payoff of a compound option is

$$\text{Co}_{T_0} = (C(S_{T_0}, \tau, K) - K_0)^+$$

where $C(S_{T_0}, \tau, K)$ stands for the value at time $T_0$ of a standard call option with strike price $K$ and expiry date $T = T_0 + \tau$. Calculate the price of a compound option at $t = 0$ when $\sigma$ and $r$ are constant.

Solution in page 74

## 5.6 At The Money Approximation ♠ (BNP)

Prove the following approximation for the price of an at the money call option

$$C \simeq 0.4S\sigma\sqrt{T}$$

Solution in page 74

## 5.7 End of Times ♠♠ (UBS)

Let $X_n$ be a sequence of positive random variables, such that $\mathbb{E}[X_n] = a$ and

$$\lim_{n \to +\infty} X_n = 0 \quad \text{a.s}$$

show that

$$\lim_{n \to +\infty} \mathbb{E}|X_n - K| = a + K$$

Can this result be applied to a financial option?

Solution in page 74

## 5.8 Connected Greeks ♠ (Deutsche Bank)

Write an approximate relationship between the theta and the gamma of a hedged option.

Solution in page 77

## 5.9 Lost in Translation ♠ (Societe Generale)

A stock is worth 100 today, there are zero interest rates. The strategist for this sector calls you and tells you that the stock can be worth 90 or 110 tomorrow, respectively with a probability $p$ and $(1 - p)$. You need to price an ATM call for a client, how do you proceed?

Solution in page 77

## 5.10 Playoff Payoff ♠ (Societe Generale)

In NBA Playoffs the teams play series of 7 games, whoever wins 4 games first wins. 2 teams $T_1$ and $T_2$ are playing a playoff series and you would like to bet on the overall winner. You would like to bet an amount B, receive 0 is you lose or 2B if you win. However the booking website only allows to bet on individual games. When you bet an amount C on an individual game on the website, your payoff is 0 if you lose or 2C if you win. Can you still achieve the desired payoff?

Solution in page 77

## 5.11 Swap That Variance ♠♠ (Hedge Fund)

How can you replicate a variance swap? Why are variance swaps more popular than volatility swaps?

Solution in page 79

## 5.12 No Future ♠ (Deutsche Bank)

What is the main difference between futures contracts and forward contracts?

Solution in page 81

## 5.13 American Dream ♠ (UBS)

Is it ever optimal to early exercise an American Call Option?

---

Solution in page 81

## 5.14 Back to The Forward ♠ (Deutsche Bank)

Prove that the price of a forward contract with strike $K$ at time 0 and expiry $T$ is

$$P = S_0 - Ke^{-rT}$$

We assume no dividends and no repo rate.

---

Solution in page 82

# Chapter 6

# Finance - Solutions

# Finance - Solutions

## 6.1 Binary Hedging - Solution

**Question :** A trader suggests the following binary hedging strategy for a call option:

- sell a call option at strike $K > S_0$
- buy the stock at $K$ when $S_t$ is increasing and crosses $K$
- sell the stock at $K$ when $S_t$ is decreasing and crosses $K$

What is wrong with this strategy?

**Solution :** This paradox is more than a simple puzzle. The question is called the stop-go paradox and was discussed in several publications (Seidenverg (1988) Carr (1989) Ingersoll (1987) El Karoui (1978)). Generally many interview candidates invoke transaction costs, liquidity or the impossibility to hit an exact price. But all these answers are incorrect because Black Scholes assumptions allow you to build this portfolio. The second type of answer is usually about the portfolio not being self financing because the trader would need to start with $K$ in cash. This is correct but could be addressed using forward contracts for example. We could also borrow the needed cash and the paradox would still be unsolved if the rates are zero.

The correct short answer is that this portfolio is not continuously derivable at $K$, this discontinuity can be crossed an infinity of times by the stochastic process, making it not self-financed.

Let us break the paradox mathematically. We construct the portfolio

$$V(t) = -\mathbb{1}_{\{S_t > KP(t)\}} KP(t) + \mathbb{1}_{\{S_t > KP(t)\}} S_t$$

where P(t) is the bond's price. This portfolio replicates the terminal payoff and it must satisfy the following equation for all $t$ in order to be self-financed

$$V(t) = V(0) + \int_0^t m(u) dP(u) + \int_0^t n(u) dS_u$$

For simplicity we take rates constant equal to zero (the general case can be reduced with a change of numeraire). The portfolio is then

$$V(t) = -\mathbb{1}_{\{S_t > K\}} K + \mathbb{1}_{\{S_t > K\}} S_t$$

and the self-financing condition becomes

$$V(t) = V(0) + \int_0^t n(u)dS_u$$

where

$$n(u) = \mathbb{1}_{\{S_u > K\}}$$

the portfolio is self-financed only if the following equation holds

$$V(t) - V(0) \stackrel{?}{=} \int_0^t \mathbb{1}_{\{S_u > K\}} dS_u$$

$$g(S_t) = \mathbb{1}_{\{S_t > K\}}(S_t - K) - (S_0 - K)^+ \stackrel{?}{=} \int_0^t \mathbb{1}_{\{S_u > K\}} dS_u$$

The key here is that $g$ is not $C^2$ and we cannot apply the usual Itô's lemma, but we can use the Tanaka's formula (see page 127) because $g$ is $C^2$ outside a finite set of points.

$$g(S_t) = g(S_0) + \int_0^t g'(S_u)dS_u + \lim_{\epsilon \to 0} \frac{1}{2\epsilon} |\{u \in [0, t]; S_u \in [K - \epsilon, K + \epsilon]\}|$$

where $g'$ is the weak derivative of $g$ and $|A|$ is the Lebesgue measure of $A$. Therefore

$$V(t) - V(0) = \int_0^t \mathbb{1}_{\{S_u > K\}} dS_u + \lim_{\epsilon \to 0} \frac{1}{2\epsilon} |\{u \in [0, t]; S_u \in [K - \epsilon, K + \epsilon]\}|$$

The last term does not converge towards zero and the portfolio is not self-financed, breaking the apparent paradox.

In real business conditions this hedging method is not used because of liquidity and the additional risk associated with this book management method. The delta hedging method is preferred, the trader accepts to pay small regular hedging costs in exchange for a much lower risk.

## 6.2 Exchange Option - Solution

**Question :** The payoff of an exchange option at expiry is

$$\text{Ex(T)} = (S_1(T) - S_2(T))^+$$

Calculate the price of an exchange option at $t = 0$ when $\rho$ is the correlation between $S_1$ and $S_2$ and $\sigma$ and $r$ are constant.

**Solution :** The dynamics of $S_1$ and $S_2$ are given by

$$dS_1(t) = rS_1(t)dt + \sigma_1 S_1(t)dw_1 \ , \ S_1(0) = s_1$$
$$dS_2(t) = rS_2(t)dt + \sigma_2 S_2(t)dw_2 \ , \ S_2(0) = s_2$$

where $w_1, w_2$ are Brownian motions with $E[dw_1 dw_2] = \rho dt$. We denote C the value of the exchange option at $t = 0$.

$$\begin{aligned} C &= e^{-rt}\mathbb{E}\left[\max\left(S_1(T) - S_2(T), 0\right)\right] \\ &= \mathbb{E}\left[\tilde{S}_2(T)\max\left(S_1(T)/S_2(T) - 1, 0\right)\right] \end{aligned}$$

where $\tilde{S}_i(t) = e^{-rt}S_i(t)$. By the Itô formula, $Y(t) = S_1(t)/S_2(t)$ satisfies

$$dY = Y\left(\sigma_2^2 - \sigma_1\sigma_2\rho\right)dt + Y\left(\sigma_1 dw_1 - \sigma_2 dw_2\right)$$

We identify a Girsanov exponential

$$\frac{1}{s_2}\tilde{S}_2(T) = \exp\left(\sigma_2 w_2(T) - \frac{1}{2}\sigma_2^2 T\right)$$

defining a change of measure

$$\frac{d\tilde{P}}{dP} = \frac{1}{s_2}S_2(T)$$

Thus

$$C = s_2\tilde{E}[\max(Y(T) - 1, 0)]$$

By the Girsanov theorem, under measure $\tilde{P}$ the process

$$d\tilde{w}_2 = dw_2 - \sigma_2 dt$$

is a Brownian motion. We can write $w_1$ as $w_1(t) = \rho w_2(t) + \sqrt{1-\rho^2}w'(t)$ where $w'(t)$ is a Brownian motion independent of $w_2(t)$ (under measure $P$). You can check that with $\tilde{P}$ defined above , $w'$ remains a Brownian motion under $P$, independent of $\tilde{w}_2$. Hence $d\tilde{w}_1$ defined by

$$\begin{aligned} d\tilde{w}_1 &= \rho d\tilde{w}_2(t) + \sqrt{1-\rho^2}dw'(t) \\ &= dw_1(t) - \rho\sigma_2 dt \end{aligned}$$

is a $\tilde{P}$ -Brownian motion. The equation for $Y$ under $\tilde{P}$ turns out-miraculously-to be

$$dY = Y\left(\sigma_1 d\tilde{w}_1 - \sigma_2 d\tilde{w}_2\right)$$

$$dY = Y\sigma dw$$

where $w$ is a standard Brownian motion and $\sigma$ is given by

$$\sigma = \sqrt{\sigma_1^2 + \sigma_2^2 - 2\rho\sigma_1\sigma_2}$$

We conclude that the exchange option is equivalent to a call option on asset $Y$ with volatility $\sigma$, strike 1 and riskless rate 0. By the Black-Scholes formula, this is

$$C(s_1, s_2) = s_1 N(d_1) - s_2 N(d_2)$$

$$d_1 = \frac{\ln(s_1/s_2) + \frac{1}{2}\sigma^2 T}{\sigma\sqrt{T}}$$
$$d_2 = d_1 - \sigma\sqrt{T}$$
$$\sigma = \sqrt{\sigma_1^2 + \sigma_2^2 - 2\rho\sigma_1\sigma_2}$$

## 6.3 Chooser Option - Solution

**Question :** A chooser option gives the right to choose at some future date $\tau$ to receive a call or put option with strike $K$ and final expiry $T > \tau$. The payoff at $\tau$ of a standard chooser option is

$$\text{Ch}(\tau) = \max\left(C(S_\tau, T-\tau, K), P(S_\tau, T-\tau, K)\right)$$

Calculate the price of a chooser option at $t = 0$ when $\sigma$ and $r$ are constant.

**Solution :** Recall that the call-put parity at $\tau$ and maturity $T$ implies that

$$P(S_\tau, T-\tau, K) = C(S_\tau, T-\tau, K) - S_\tau + Ke^{-r(T-\tau)}$$

We can rewrite the chooser option value at $\tau$

$$\text{Ch}(\tau) = \max\left\{C(S_\tau, T-\tau, K), C(S_\tau, T-\tau, K) - S_\tau + Ke^{-r(T-\tau)}\right\}$$

or finally

$$\text{Ch}(\tau) = C(S_\tau, T-\tau, K) + \left(Ke^{-r(T-\tau)} - S_\tau\right)^+$$

The last equality implies immediately that the standard chooser option is equivalent to the portfolio composed of a long call option and a long put option (with different exercise prices and different expiry dates), so that its arbitrage price equals, for every $t \in [0, \tau]$,

$$\text{Ch}(t) = C(S_t, T-t, K) + P\left(S_t, \tau - t, Ke^{-r(T-\tau)}\right)$$

In particular, using the Black-Scholes formula, we get for $t = 0$

$$\text{Ch}(0) = S_0\left(N(d_1) - N\left(-\bar{d}_1\right)\right) + Ke^{-rT}\left(N\left(-\bar{d}_2\right) - N(d_2)\right)$$

where

$$d_{1,2} = \frac{\ln(S_0/K) + \left(r \pm \frac{1}{2}\sigma^2\right)T}{\sigma\sqrt{T}}$$

and

$$\bar{d}_{1,2} = \frac{\ln(S_0/K) + rT \pm \frac{1}{2}\sigma^2\tau}{\sigma\sqrt{\tau}}$$

## 6.4 Forward Start Option - Solution

**Question :** The terminal payoff of a forward start call option is

$$\text{Fs}(T) = (S_T - KS_{T_0})^+$$

where $0 < T_0 < T$. Calculate the price of a forward start call option at $t = 0$ when $\sigma$ and $r$ are constant.

**Solution :** Let us consider the case of a forward-start call option, with terminal payoff

$$\text{Fs}(T) = (S_T - KS_{T_0})^+$$

To find the price at time $t \in [0, T_0]$ of such an option, it suffices to consider its value at the delivery date $T_0$, that is, the price at time $T_0$ of the at-the-money option with expiry date $T$. Thus, we have

$$\text{Fs}(T_0) = C\left(S_{T_0}, T - T_0, KS_{T_0}\right)$$

Where $C(S, T, K)$ is the price of call option with spot $S$, time to expiry $(T - T_0)$ and strike $K$. We can factor $S_{T_0}$ and divide the spot and the strike

$$C\left(S_{T_0}, T - T_0, KS_{T_0}\right) = S_{T_0} C\left(1, T - T_0, K\right)$$

since $C\left(1, T - T_0, K\right)$ is non-random, the option's value at time 0 equals

$$\text{Fs}(T_0) = \mathbb{E}_0(S_{T_0}) \exp(-rT_0) C\left(1, T - T_0, K\right) = C\left(S_0, T - T_0, S_0 K\right)$$

$$\text{Fs}(T_0) = S_0 C\left(1, T - T_0, K\right) = C\left(S_0, T - T_0, S_0 K\right)$$

Note that the Forward Start Option has a closed formula when $\sigma$ is constant, but its pricing becomes much more complex when the volatility surface is not trivial. Forward Start Option are notoriously sensitive to the forward skew and require a specific model, for example a stochastic volatility model.

## 6.5 Compound Option - Solution

**Question :** The payoff of a compound option is

$$\text{Co}_{T_0} = \left(C\left(S_{T_0}, \tau, K\right) - K_0\right)^+$$

where $C\left(S_{T_0}, \tau, K\right)$ stands for the value at time $T_0$ of a standard call option with strike price $K$ and expiry date $T = T_0 + \tau$. Calculate the price of a compound option at $t = 0$ when $\sigma$ and $r$ are constant.

**Solution :** There is no simple closed formula for Compound Option, the final result will be an integral. We start with the price of a call option

$$C(s, \tau, K) = sN\left(d_1(s, \tau, K)\right) - Ke^{-r\tau}N\left(d_2(s, \tau, K)\right)$$

Moreover, since under $\mathbb{P}^*$ we have

$$S_{T_0} = S_0 \exp\left(\sigma\sqrt{T_0}\xi + \left(r - \frac{1}{2}\sigma^2\right)T_0\right)$$

where $\xi$ has a standard Gaussian probability law under $\mathbb{P}^*$, the price of the compound option at time 0 equals

$$\text{Co}_0 = e^{-rT_0}\int_{x_0}^{\infty}\left(g(x)N\left(\hat{d}_1\right) - Ke^{-r\tau}N\left(\hat{d}_2\right) - K_0\right)n(x)dx$$

where $\hat{d}_i = d_i(g(x), \tau, K)$ for $i = 1, 2$, $n$ is the density of the standard normal distribution, the function $g : \mathbb{R} \to \mathbb{R}$ is given by the formula

$$g(x) = S_0 \exp\left(\sigma\sqrt{T_0}x + \left(r - \frac{1}{2}\sigma^2\right)T_0\right)$$

and, finally, the constant $x_0$ is defined implicitly by the equation

$$x_0 = \inf\left\{x \in \mathbb{R} | C(g(x), \tau, K) \geq K_0\right\}$$

Straightforward calculations yield

$$d_1(g(x), \tau, K) = \frac{\ln\left(S_0/K\right) + \sigma\sqrt{T_0}x + rT - \sigma^2 T_0 + \frac{1}{2}\sigma^2 T}{\sigma\sqrt{T - T_0}}$$

and

$$d_2(g(x), \tau, K) = \frac{\ln\left(S_0/K\right) + \sigma\sqrt{T_0}x + rT - \frac{1}{2}\sigma^2 T}{\sigma\sqrt{T - T_0}}$$

## 6.6 At The Money Approximation - Solution

**Question :** Prove the following approximation for the price of an at the money call option

$$C \simeq 0.4S\sigma\sqrt{T}$$

**Solution :** We start with the call option price formula

$$C = S\phi(d_1) - Ke^{-rT}\phi(d_2)$$

where

$$d_1 = \frac{ln\left(\frac{S_0}{K}\right) + rT + \frac{\sigma^2 T}{2}}{\sigma\sqrt{T}} \ , \ d_2 = \frac{ln\left(\frac{S_0}{K}\right) + rT - \frac{\sigma^2 T}{2}}{\sigma\sqrt{T}}$$

the call is at the money so

$$C = S\left(\phi(d_1) - \phi(d_2)\right)$$

where

$$d_1 = \frac{\sigma\sqrt{T}}{2} \ , \ d_2 = \frac{-\sigma\sqrt{T}}{2}$$

and

$$\phi(d_1) - \phi(d_2) = \int_{d_2}^{d_1} n(x)dx$$

where $n$ is the density function of the standard normal distribution. Now the volatility is usually below 0.5, and for relatively short expiries (less than 2 years) we can approximate the integral with the area of a rectangle

$$\int_{d_2}^{d_1} n(x)dx = \int_{-d_1}^{d_1} n(x)dx \simeq 2n(0)d_1 \simeq 0.4\sigma\sqrt{T}$$

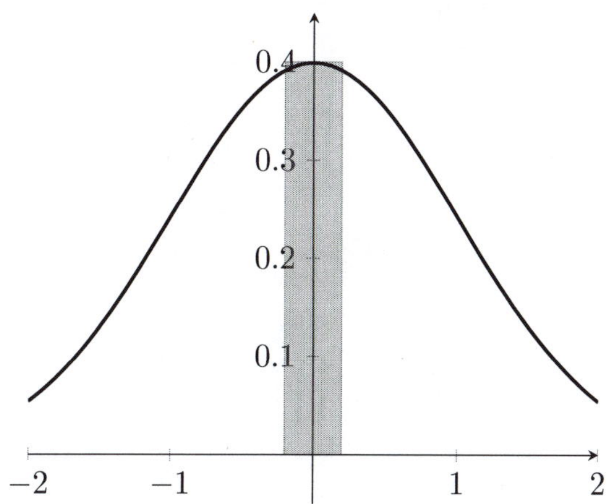

and

$$C \simeq 0.4S\sigma\sqrt{T}$$

## 6.7 End of Times - Solution

**Question :** Let $X_n$ be a sequence of positive random variables, such that $\mathbb{E}[X_n] = a$ and

$$\lim_{n\to+\infty} X_n = 0 \text{ a.s}$$

show that

$$\lim_{n\to+\infty} \mathbb{E}|X_n - K| = a + K$$

Can this result be applied to a financial option?

**Solution :** To solve elegantly this question, a formula to remember is

$$|X_n - K| = X_n + K - 2\min(Xn, K)$$

$X_n$ converges a.s towards 0, therefore $\mathbb{E}[\min(Xn, K)]$ converges towards 0 too. So

$$lim_{n\to+\infty}E(|X_n - K|) = lim_{n\to+\infty}\mathbb{E}[X_n] + K = a + K$$

The financial option we can associate with this result is a straddle with payoff

$$\text{Payoff}_T = |S_T - K|$$

At the money forward the price becomes

$$P = \mathbb{E}\left[\exp(-rT)|S_T - K|\right] = \mathbb{E}\left[|\exp(-rT)S_T - S_0|\right]$$

$$P = \mathbb{E}\left[|S_0 Y_T - S_0|\right]$$

where

$$Y_T = \exp\left(\sigma\sqrt{T}\xi - \frac{1}{2}\sigma^2 T\right)$$

with $\xi$ a standard normal variable. Now let A be a real number, we have

$$P\left(\sigma\sqrt{T}\xi - \frac{1}{2}\sigma^2 T > A\right) = P\left(\xi > \frac{A}{\sigma\sqrt{T}} + \frac{\sigma\sqrt{T}}{2}\right)$$

We see that if $\sigma \to +\infty$ or $T \to +\infty$, $\left(\sigma\sqrt{T}\xi - \frac{1}{2}\sigma^2 T\right) \to +\infty$ a.s and $Y_n \to 0$ a.s. Using the previous result, we conclude that the price of the at the money forward straddle price converges towards $2S_0$ when the maturity is very long or the volatility very high.

## 6.8 Connected Greeks - Solution

**Question :** Write an approximate relationship between the theta and the gamma of a hedged option.

**Solution :** Let $\pi(t)$ the value of the hedged option position. The change in the portfolio value is

$$d\pi = \text{PnL} = V_t dt + V_S dS + \frac{1}{2} V_{SS} \sigma^2 S^2 dt$$

$V_S$ is the delta, equal to zero for a hedged portfolio, $V_{SS}$ is the $\Gamma$. In average, the PnL should be zero

$$\mathbb{E}(\text{PnL}) \simeq 0 \simeq \Theta dt + \frac{1}{2} \Gamma \sigma^2 S^2 dt$$

We have in average the relationship

$$\Theta \simeq -\frac{1}{2} \Gamma \sigma^2 S^2$$

## 6.9 Lost in Translation - Solution

**Question :** A stock is worth 100 today, there are zero interest rates. The strategist for this sector calls you and tells you that the stock can be worth 90 or 110 tomorrow, respectively with a probability $p$ and $(1-p)$. You need to price an ATM call for a client, how do you proceed?

**Solution :** This is a classic trick question where the probability of moving can be provided by a colleague, a friend, an insider, or sometimes God himself. You should reject the given probability $p$ because you want to price your option in the risk neutral probability. In the risk neutral probability $q$ the forward is defined by the information about the interest rates. There are zero interest rates so

$$\mathbb{E}(S_{t+1}) = S_t = q S_d + (1-q) S_u$$

So $q = 0.5$ and

$$\text{Price}_{\text{Call}} = 10\frac{1}{2} + 0\frac{1}{2} = 5$$

## 6.10 Playoff Payoff - Solution

**Question :** In NBA Playoffs the teams play series of 7 games, whoever wins 4 games first wins. 2 teams $T_1$ and $T_2$ are playing a playoff series and you would like to bet on the overall winner. You would like to bet an amount B, receive 0 is you lose or 2B if you win. However the booking website only allows to bet on individual games. When you bet an amount C on an individual game on the website, your payoff is 0 if you lose or 2C if you win. Can you still achieve the desired payoff?

**Solution :** This is a replication question. You are asked to replicate a 7 days digital options on the number of wins of a team, using 1 day small digital options. Our advice for this type of questions is to work out the full tree, as the existence of the replication is not guaranteed, and if the interviewer makes a mistake you might be in a case where the replication is impossible. By detailing the tree you will be able to check that the replication is possible and find the betting strategy. We create the tree, starting from the desired payoff, and compute it backwards. We denote at each step $p$ the amount of money in the player's pocket and $b$ the amount he will bet at the next game. We go up in the tree when $T_1$ wins the game. At every node we solve

$$p_u(t) = p(t-1) + b$$
$$p_d(t) = p(t-1) - b$$

which means

$$p(t-1) = \frac{p_u(t) + p_d(t)}{2}$$
$$b = \frac{p_u(t) - p_d(t)}{2}$$

Starting tree

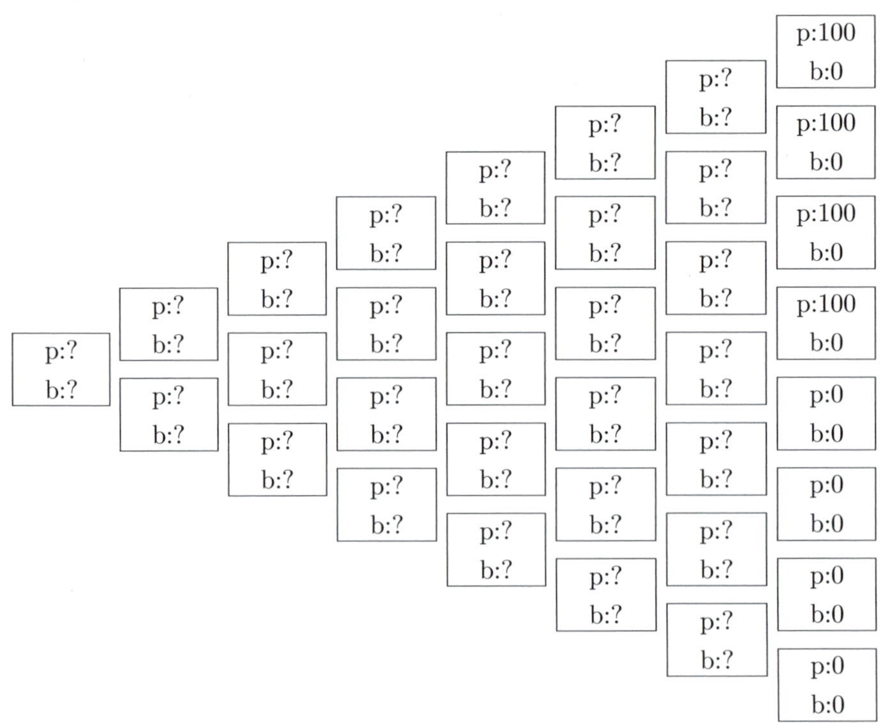

Populated tree

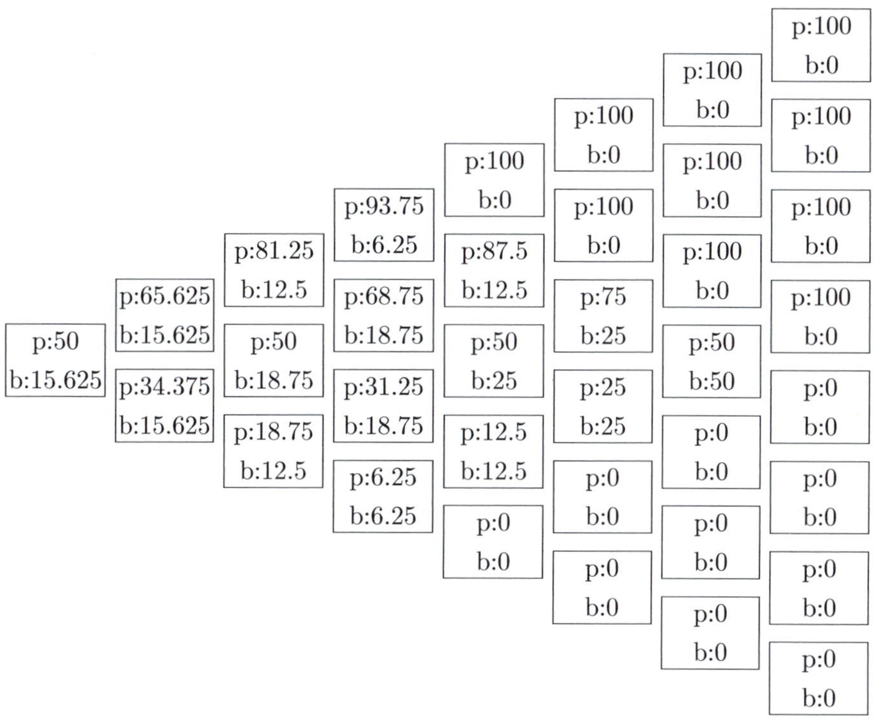


We have created a betting strategy for a starting bet of 50, yielding a payoff of 100 if $T_1$ wins 4 games or more, 0 otherwise. In order to bet B and receive 2B when $T_1$ wins, we can multiply all the values in the tree by $\frac{B}{50}$.

## 6.11 Swap That Variance - Solution

**Question :** How can you replicate a variance swap? Why are variance swaps more popular than volatility swaps?

**Solution :** Related work on this question dates back as far as Neuberger's paper (1990). Variance swaps pay the difference between the future realized variance of the price changes of the underlying, and some prespecified strike price that we label as $K_{\text{var.}}$ Its fair strike at inception is determined by the implied volatility skew, but its final payout is a function of the realized variance. The payoff of the variance swap at expiry is given by

$$\text{Payoff}_{\text{ Variance Swap}} = \left(\text{RV}(T) - K_{\text{var}}^2\right)$$

where $\mathrm{RV}(T)$ is the realized variance

$$\mathrm{RV}(T) = \frac{A}{N_{\mathrm{E}}} \sum_{i=1}^{N_{\mathrm{A}}} \ln^2 \left( \frac{S(t_i)}{S(t_{i-1})} \right)$$

where $N_{\mathrm{E}}$ is the expected number of sampling points given by the number of trading days; and $N_{\mathrm{A}}$ is the actual number of trading days where $0 = t_0, t_1, \ldots, t_{N_{\mathrm{A}}} = T$ between time $t = 0$ and $T$. The idea for the replication is to consider the stock equation

$$S_T = S_0 \exp\left( rt - \frac{\int_0^T \sigma^2 u du}{2} + \sigma W_T \right)$$

We take the log

$$\log(S_T) = \log(S_0) + rT - \frac{\int_0^T \sigma^2 u du}{2} + \sigma W_T$$

and the expectation

$$\mathbb{E}\left(\log(S_T)\right) = \log(S_0) + rT - \mathbb{E}\left( \frac{\int_0^T \sigma^2 u du}{2} \right)$$

and we find

$$\mathbb{E}\left( \frac{\int_0^T \sigma^2 u du}{T} \right) = \frac{2}{T}\left( -\mathbb{E}\left(\log(S_T)\right) + \log(S_0) + rT \right)$$

Having expressed the payoff as a function of the final value of the stock $S_T$, we are guaranteed that a vanilla options replication portfolio exists. The Breeden Litzengerger formula (see page 126) shows that

$$\frac{\partial^2 C}{\partial K^2} = \frac{\partial^2 P}{\partial K^2} = e^{-rT} p(k)$$

where $p$ is the probability density of $S_T$. So any payoff of the form $g(S_T)$ can be replicated with calls or puts

$$\text{Price} = \exp(-rT) \int_0^{+\infty} g(x) p(x) dx$$

In practice the integral is approximated as a sum

$$\text{Price} \simeq \exp(-rT) \sum_{i=0}^{+\infty} g(x_i) p(x_i) \Delta = \sum_{i=0}^{+\infty} g(x_i) \frac{\partial^2 C}{\partial K^2}(x_i) \Delta$$

where $\Delta = x_{i+1} - x_i$ and the second derivative of the call options is obtained as

$$\frac{\partial^2 C}{\partial K^2}(x_i) \simeq \frac{1}{\Delta^2}\left( C(x_i - \Delta) - 2C(x_i) + C(x_i + \Delta) \right)$$

The demand for variance swaps is greater because variance swaps are a much more natural hedging instrument. Taking the example of the Delta-hedged call, it is the variance that appears in the P&L formula, not the volatility. The variance is scaled by time, giving it an additive property, whereas volatility, which is the square root of variance, loses this property of linearity. Another reason, now for the seller of a variance or volatility swap, is that there exist robust replication strategies for variance swaps.

## 6.12 No Future - Solution

**Question :** What is the main difference between futures contracts and forward contracts?

**Solution :** This question is a classic, and although it is more frequent for trader positions, it is also a question we see in quant interviews. The main difference between forward and futures contracts is that futures contracts are traded on exchanges and forwards are traded over-the-counter. Because of this distinction, you can only trade specific futures contracts that are traded on the exchange. Forward contracts are more flexible because they are privately negotiated, and can represent any assets and can change settlement dates should both parties agree.

## 6.13 American Dream - Solution

**Question :** Is it ever optimal to early exercise an American Call Option?

**Solution :** It is worth having a fresh look at this classic question, given the recent negative interest rates wave. The usual expected answer is: it is never optimal to early exercise an American Call Option if there is no dividends. But this answer is assuming positive interest rates.

Let us start by assuming no dividends and positive interest rates. We consider a call option with strike $K$ on a stock with spot value $S_t$. If $S_t < K$ the option is out of the money, the payoff today would be zero and it is clearly better to wait as we could expect a better payoff. if $S_t > K$ the option holder shall compare the exercise value $V_e$ to the value of keeping ($V_k$) the option until a future time $T$. We denote $\tau = T - t$

$$V_e = S_t - K$$

$$V_k = e^{-r\tau}\mathbb{E}((S_T - K)^+)$$

Now $(S_T - K)^+ > S_T - K$ so

$$V_k > e^{-r\tau}\mathbb{E}(S_T - K) = e^{-r\tau} S_t e^{r\tau} - Ke^{-r\tau} = S_t - Ke^{-r\tau} > S_T - K = V_e$$

and it is not optimal to exercise the option early.

We assume now that the interest rates are negative. If $S_t < K$ the option is out of the money, the payoff today would be zero and it is still clearly better to wait. If the option is deep in the money ($S_t >> K$) then the probability to be out of the money at time $T$ becomes very small and we can approximate $V_k$

$$V_e = S_t - K$$

$$V_k = e^{-r\tau}\mathbb{E}((S_T - K)^+) \simeq e^{-r\tau}\mathbb{E}((S_T - K)) = S_t - Ke^{-r\tau} < S_t - K = V_e$$

We see a clear pattern: it is better to keep the option for low spots and to exercise it for high spots. We can visualize it by drawing $V_e$ and $V_k$. Remember that $V_e$ is simply the intrinsic value of a call and $V_k$ is a European call option with maturity $\tau$. In the figure below $S_c$ represents the critical spot where exercise becomes optimal.

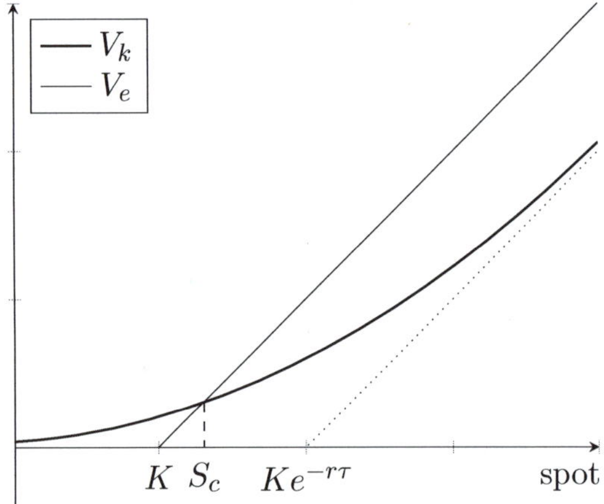


When we add dividends, depending on the expected level of dividends, any rates scenario can become an optimal exercise case. If a dividend $d$ is expected at some date $t_d$, the spot is expected to drop of that amount. Depending on spot level and dividend level it can become optimal to exercise just before the drop.

In conclusion, it can be optimal to exercise an American Call Option in the following cases

- If there are dividends
- If the interest rates are negative

## 6.14 Back to The Forward - Solution

**Question :** Prove that the price of a forward contract with strike $K$ at time 0 and expiry $T$ is

$$P = S_0 - Ke^{-rT}$$

We assume no dividends and no repo rate.

**Solution :** This another typical question for trading roles question, but we see it sometimes for quant roles. We can establish the price of a forward using arbitrage arguments.

We assume first that the price of the forward is higher than the suggested value

$$P > S_0 - Ke^{-rT}$$

We construct the following portfolio: at time 0 sell the forward and buy the stock

- sell the forward and receive $P > S_0 - Ke^{-rT}$
- buy the stock and spend $S_0$
- we borrow $Ke^{-rT}$

The cash position at time zero is $P - S_0 + Ke^{-rT} > 0$. At expiry

- we deliver the stock to the client
- we receive $K$ from the client
- we pay back the borrowed money with interest, we pay $K$

We have locked a gain $P - S_0 + Ke^{-rT} > 0$ with certainty and constructed an arbitrage.

We assume next that the price of the forward is lower than the suggested value

$$P < S_0 - Ke^{-rT}$$

We construct the following portfolio: at time 0 buy the forward and short the stock

- buy the forward and pay $P < S_0 - Ke^{-rT}$
- short the stock and receive $S_0$
- we invest $Ke^{-rT}$

The cash position at time zero is $-P + S_0 - Ke^{-rT} > 0$. At expiry

- the counterparty delivers the stock, filling our short position
- we pay $K$ to the client
- we take back the money invested at the risk free rate $K$

We have locked a gain $-P + S_0 - Ke^{-rT} > 0$ with certainty and constructed an arbitrage. So in conclusion the price of the forward has to be

$$P = S_0 - Ke^{-rT}$$

## Chapter 7

# Programming

# Programming

## 7.1 Best Sort ♠♠ (Hedge Fund)

For a comparison sort (where nothing is assumed about the items except that they can be compared), what is the best complexity?

Prove that this is the tightest bound.

Solution in page 91

## 7.2 Merge Sort ♠ (Citi)

Write a code for merge sort in the language of your choice or pseudo code.

Solution in page 91

## 7.3 Quick Sort ♠ (Citi)

Write a code for quick sort in the language of your choice or pseudo language.

Solution in page 92

## 7.4 Struct vs Class ♠ (Credit Suisse)

how does a struct differ from a class in C++ ?

Solution in page 93

## 7.5 Friend Class ♠ (Credit Suisse)

What is a friend class in C++ ?

Solution in page 94

## 7.6 Singleton ♠♠♠ (Goldman Sachs)

Explain C++ Singleton design pattern. Implement a version in C++ .

Solution in page 94

## 7.7 Is Python Compiled? ♠♠ (Goldman Sachs)

Is python a compiled language?

---

Solution in page 95

## 7.8 Python Hash ♠♠ (Goldman Sachs)

How is the hash function used in python?

---

Solution in page 95

## 7.9 Python Self ♠♠ (Hedge Fund)

Explain the keyword self in python.

---

Solution in page 97

## 7.10 Python Mutable ♠ (Hedge Fund)

What are mutable and immutable data types in Python?

---

Solution in page 97

## 7.11 Virtual Reality ♠ (BNP)

What is a virtual function in C++ ?

---

Solution in page 98

Chapter 8

# Programming - Solutions

# Programming - Solutions

## 8.1 Best Sort - Solution

**Question :** For a comparison sort (where nothing is assumed about the items except that they can be compared), what is the best complexity?

Prove that this is the tightest bound.

**Solution :** The best complexity in this case is $\mathcal{O}(N\log(N))$. Let $N$ be the number of elements in the list. There are $N!$ possible permutations of the list. Sorting the list is equivalent to discovering which of these permutations sorts the list. By symmetry a given comparison eliminates half of the remaining permutations, because the number of permutations where $a_i$ is after $a_j$ is equal to the number of permutations where $a_j$ is after $a_i$. Therefore the best complexity $T$ satisfies

$$\frac{N!}{2^T} = 1$$

and

$$T = \frac{\log(N!)}{\log(2)}$$

Note that $\log(N!) \simeq N\log(N)$, we can prove it with the Sterling formula, or by decomposing the log and seeing a Riemann integral

$$\log(N!) = \log(2) + \log(3) + \cdots + \log(N) \simeq \int_1^N \log(u)du = N\log(N) - N$$

Therefore the best complexity is the highest order and

$$T \simeq \mathcal{O}(N\log(N))$$

## 8.2 Merge Sort - Solution

**Question :** Write a code for merge sort in the language of your choice or pseudo code.

**Solution :** Our advice for this type of question is to use python. If you choose C++ you will make the task longer and if you choose pseudo code you will most likely raise questions about which subfunctions you assumed to be available. By using python you will write short code and tick the python box with the interviewer.

```
# Python program for implementation of MergeSort
def mergeSort(arr):
```

```
if len(arr) >1:
    mid = len(arr)//2 # Finding the mid of the array
    L = arr[:mid] # Dividing the array elements
    R = arr[mid:] # into 2 halves

    mergeSort(L) # Sorting the first half
    mergeSort(R) # Sorting the second half

    i = j = k = 0

    # Copy data to temp arrays L[] and R[]
    while i < len(L) and j < len(R):
        if L[i] < R[j]:
            arr[k] = L[i]
            i+= 1
        else:
            arr[k] = R[j]
            j+= 1
        k+= 1

    # Checking if any element was left
    while i < len(L):
        arr[k] = L[i]
        i+= 1
        k+= 1

    while j < len(R):
        arr[k] = R[j]
        j+= 1
        k+= 1
```

## 8.3 Quick Sort - Solution

**Question :** Write a code for quick sort in the language of your choice or pseudo language.

**Solution :**

```
# Python program for implementation of Quicksort Sort
# This function takes last element as pivot, places
# the pivot element at its correct position in sorted
# array, and places all smaller (smaller than pivot)
# to left of pivot and all greater elements to right
# of pivot
```

```python
def partition(arr,low,high):
    i = ( low-1 )          # index of smaller element
    pivot = arr[high]      # pivot

    for j in range(low , high):

        # If current element is smaller than or
        # equal to pivot
        if   arr[j] <= pivot:

            # increment index of smaller element
            i = i+1
            arr[i],arr[j] = arr[j],arr[i]

    arr[i+1],arr[high] = arr[high],arr[i+1]
    return ( i+1 )

# The main function that implements QuickSort
# arr[] --> Array to be sorted,
# low  --> Starting index,
# high  --> Ending index

# Util Function to do Quick sort
def quickSort_Util(arr,low,high):
    if low < high:

        # pi is partitioning index, arr[p] is now
        # at right place
        pi = partition(arr,low,high)

        # Separately sort elements before
        # partition and after partition
        quickSort_Util(arr, low, pi-1)
        quickSort_Util(arr, pi+1, high)

# Main Function to do Quick sort
def quickSort(arr):
    quickSort_Util(arr, 0, len(arr)-1)
```

## 8.4 Struct vs Class - Solution

**Question :** how does a struct differ from a class in C++ ?

**Solution :** The only difference between a struct and class in C++ is the default accessibility of member variables and methods. In a struct they are public; in a class they are private.

## 8.5 Friend Class - Solution

**Question :** What is a friend class in C++ ?

**Solution :** A friend class can access private and protected members of other class in which it is declared as friend. It is sometimes useful to allow a particular class to access private members of other class.

## 8.6 Singleton - Solution

**Question :** Explain C++ Singleton design pattern. Implement a version in C++ .

**Solution :** Singleton design pattern is a software design principle that is used to restrict the instantiation of a class to one object. This is useful when exactly one object is needed to coordinate actions across the system. For example, if you are using a logger, that writes logs to a file, you can use a singleton class to create such a logger. You can create a singleton class using the following code

```
#include <iostream>

using namespace std;

class Singleton {
   static Singleton *instance;
   int data;

   // Private constructor so that no objects can be created.
   Singleton() {
      data = 0;
   }

   public:
   static Singleton *getInstance() {
      if (!instance)
      instance = new Singleton;
      return instance;
   }

   int getData() {
      return this -> data;
```

```
    }

    void setData(int data) {
        this -> data = data;
    }
};

//Initialize pointer to zero so that it can be initialized
//in first call to getInstance
Singleton *Singleton::instance = 0;

int main(){
    Singleton *s = s->getInstance();
    cout << s->getData() << endl;
    s->setData(100);
    cout << s->getData() << endl;
    return 0;
}
```

This will give the output

0
100

## 8.7 Is Python Compiled? - Solution

**Question :** Is python a compiled language?

**Solution :** This is a common question. Usually, "compile" means to convert a program in a high-level language into a binary executable full of machine code (CPU instructions). In Python, the source code is compiled into a much simpler form called bytecode. These are instructions similar in spirit to CPU instructions, but instead of being executed by the CPU, they are executed by software called a virtual machine. So the answer is Python is not directly compiled to CPU instructions, but it is still compiled to virtual machine language.

## 8.8 Python Hash - Solution

**Question :** How is the hash function used in python?

**Solution :** Hash tables are used to implement map and set data structures in many common programming languages, such as C++ , Java, and Python. Python uses hash tables for dictionaries and sets. A hash table is an unordered collection of key-value pairs, where each key is unique. Hash tables offer a combination of efficient lookup, insert and delete operations. These are the best properties of arrays and linked lists.

Hashing is the process of using an algorithm to map data of any size to a fixed length. This is called a hash value. Hashing is used to create high performance, direct access data structures where large amount of data is to be stored and accessed quickly. Hash values are computed with hash functions.

An object is hashable if it has a hash value which never changes during its lifetime. (It can have different values during multiple invocations of Python programs.) A hashable object needs a __hash__() method. In order to perform comparisons, a hashable needs an __eq__() method.

Note: Hashable objects which compare equal must have the same hash value. Hashability makes an object usable as a dictionary key and a set member, because these data structures use the hash value internally. Python immutable built-in objects are hashable; mutable containers (such as lists or dictionaries) are not. Objects which are instances of user-defined classes are hashable by default. They all compare unequal (except with themselves), and their hash value is derived from their id().

Note: If a class does not define an __eq__() method it should not define a __hash__() operation either; if it defines __eq__() but not __hash__(), its instances will not be usable as items in hashable collections. Python hash() function The hash() function returns the hash value of the object if it has one. Hash values are integers. They are used to quickly compare dictionary keys during a dictionary lookup. Objects can implement the __hash__() method.

Python immutable builtins, such as integers, strings, or tuples, are hashable. Below a class implementation with a hash function.

```
class User:

    def __init__(self, name, job):

        self.name = name
        self.job = job

    def __eq__(self, other):

        return self.name == other.name \
            and self.job == other.job

    def __str__(self):
        return f'{self.name} {self.job}'
```

## 8.9 Python Self - Solution

**Question :** Explain the keyword self in python.

**Solution :**

self represents the instance of the class. By using the "self" keyword we can access the attributes and methods of the class in python. It binds the attributes with the given arguments.

The reason you need to use self is because Python does not use the @ syntax to refer to instance attributes. Python decided to do methods in a way that makes the instance to which the method belongs be passed automatically, but not received automatically: the first parameter of methods is the instance the method is called on.

```
class car():

    # init method or constructor
    def __init__(self, model, color):
        self.make = make
        self.color = color

    def show(self):
        print("make is", self.make )
        print("color is", self.color )

# both objects have different self which
# contain their attributes
audi = car("audi", "blue")
ferrari = car("ferrari", "green")

audi.show()     # same output as car.show(audi)
ferrari.show()  # same output as car.show(ferrari)

# Behind the scene, in every instance method
# call, python sends the instances also with
# that method call like car.show(audi)
```

self is parameter in function and user can use another parameter name in place of it.But it is advisable to use self because it increase the readability of code.

## 8.10 Python Mutable - Solution

**Question :** What are mutable and immutable data types in Python?

**Solution :** Everything in python is an object. The objects which cannot be changed once created is called immutable type and the objects which can be changed after creation are called mutable type.

- Mutable data types: list, dict, set, byte array
- Immutable data types: int, float, complex, string, tuple, frozen set, bytes

## 8.11 Virtual Reality - Solution

**Question :** What is a virtual function in C++ ?

**Solution :** Virtual functions allow the programmer to defer the implementation of a function declared in a base class to an inherited class. For example, suppose we have a Polygon class, we can make the Area method virtual. We then inherit Squares and Triangles from the Polygon class. The Area method is then defined differently for squares and triangles. When we manipulate a Polygon object we do not need to know which sort of polygon it is, the right Area function is called automatically.

# Chapter 9

# Classic Calculations

# Classic Calculations

## 9.1 Call Option ♠♠ (Societe Generale)

Derive the formula for the price of a call option using Girsanov theorem.

---

Solution in page 105

## 9.2 Greeks ♠♠ (Societe Generale)

Calculate the greeks $\Delta$, $\Gamma$, $\mathcal{V}$, $\rho$, $\Theta$ for a call option.

---

Solution in page 107

## 9.3 Ornstein Uhlenbeck ♠♠ (JP Morgan)

Derive the formula of an Ornstein Uhlenbeck process. Calculate its expectation and variance.

---

Solution in page 108

## 9.4 Hybrid Vasicek ♠♠♠ (JP Morgan)

Derive the relationship between the stock volatility and the rates volatility in a hybrid Vasicek model.

---

Solution in page 109

## 9.5 Fokker-Planck ♠♠♠ (Morgan Stanley)

Derive the Fokker-Planck formula.

---

Solution in page 111

## 9.6 Breeden-Litzenberger ♠♠♠ (Morgan Stanley)

Derive the Breeden-Litzenberger Formula.

---

Solution in page 113

## 9.7 Local Volatility ♠♠♠ (Morgan Stanley)

Derive the Dupire Formula or Local Volatility.

---

Solution in page 114

## 9.8 Black Scholes Equation ♠♠ (BNP)

Derive the Black Scholes equation.

---

Solution in page 115

## 9.9 Black Scholes Robustness ♠♠♠ (JP Morgan)

You sell a European option for which you estimated the volatility to be $\sigma_t$. What is the PnL error if you hedge this option until expiry and that the realized volatility turns out to be $\sigma_r \neq \sigma_t$?

---

Solution in page 116

## 9.10 Local Variance as an Expectation ♠♠♠ (Natixis)

Show that the local variance can be seen as a conditional expectation of the instantaneous variance.

---

Solution in page 117

## Chapter 10

# Classic Calculations - Solutions

# Classic Calculations - Solutions

## 10.1 Call Option - Solution

**Question :** Derive the formula for the price of a call option using Girsanov theorem.

**Solution :** We denote $C$ the price at $t = 0$ of a call option.

$$C = e^{-rT}\mathbb{E}(S_T - K)^+ = e^{-rT}\mathbb{E}\Big(S_T\mathbb{1}_{(S_T \geq K)} - K\mathbb{1}_{(S_T \geq K)}\Big)$$

Where $S_t$ is the process

$$S_t = S_0 exp\Big(rt - \frac{\sigma^2 t}{2} + \sigma W_t\Big)$$

We decompose $C = A - B$ and start calculating the second term $B$

$$B = e^{-rT}K\mathbb{E}(\mathbb{1}_{(S_T \geq K)}) = e^{-rT}KP(S_T \geq K)$$

$$B = e^{-rT}KP\Big(ln\Big(\frac{S_0}{K}\Big) + rT - \frac{\sigma^2 T}{2} + \sigma W_T \geq 0\Big)$$

$$B = e^{-rT}KP\Bigg(W_T \geq \frac{ln\Big(\frac{K}{S_0}\Big) - rT + \frac{\sigma^2 T}{2}}{\sigma}\Bigg)$$

$$B = e^{-rT}KP\Bigg(W_T \leq \frac{ln\Big(\frac{S_0}{K}\Big) + rT - \frac{\sigma^2 T}{2}}{\sigma}\Bigg)$$

$$B = e^{-rT}KP\Bigg(X \leq \frac{ln\Big(\frac{S_0}{K}\Big) + rT - \frac{\sigma^2 T}{2}}{\sigma\sqrt{T}}\Bigg) = Ke^{-rT}\phi(d_2)$$

where $X \sim \mathcal{N}(0,1)$ and $\phi$ is the standard cumulative normal distribution. The calculation for $A$ is trickier

$$A = e^{-rT}\mathbb{E}(S_T\mathbb{1}_{(S_T \geq K)}) = S_0\mathbb{E}\Big(exp\Big(-\frac{\sigma^2 t}{2} + \sigma W_t\Big)\mathbb{1}_{(S_T \geq K)}\Big)$$

We identify a Girsanov change of measure where

$$Q(E) = \int_E Z_t dP$$

and

$$Z_t = exp\left( \int_0^t \sigma dWs - \int_0^t \frac{\sigma^2}{2} ds \right)$$

$$A = S_0 Q(S_T \geq K)$$

In this new measure $\hat{W}_t$ is a Brownian Motion where

$$\hat{W}_t = W_t - \int_0^t \sigma ds$$

Therefore the dynamics of $S_t$ in the new measure are

$$S_t = S_0 exp\left(rt + \frac{\sigma^2 t}{2} + \sigma \hat{W}_t\right)$$

$$A = S_0 Q\left( X \leq \frac{ln\left(\frac{S_0}{K}\right) + rT + \frac{\sigma^2 T}{2}}{\sigma\sqrt{T}} \right) = S_0 \phi(d_1)$$

We combine them to find the price of the call option

$$C = S_0 \phi(d_1) - Ke^{-rT}\phi(d_2)$$

where

$$d_1 = \frac{ln\left(\frac{S_0}{K}\right) + rT + \frac{\sigma^2 T}{2}}{\sigma\sqrt{T}}, \; d_2 = \frac{ln\left(\frac{S_0}{K}\right) + rT - \frac{\sigma^2 T}{2}}{\sigma\sqrt{T}}$$

Note that with dividends the formula becomes

$$C = S_0 e^{-qT}\phi(d_1) - Ke^{-rT}\phi(d_2)$$

$$d_1 = \frac{ln\left(\frac{S_0}{K}\right) + (r-q)T + \frac{\sigma^2 T}{2}}{\sigma\sqrt{T}}, \; d_2 = \frac{ln\left(\frac{S_0}{K}\right) + (r-q)T - \frac{\sigma^2 T}{2}}{\sigma\sqrt{T}}$$

Sometimes a different formula can be found using the forward $F = S_0 e^{(r-q)T}$

$$C = Fe^{-rT}\phi(d_1) - Ke^{-rT}\phi(d_2)$$

$$d_1 = \frac{ln\left(\frac{F}{K}\right) + \frac{\sigma^2 T}{2}}{\sigma\sqrt{T}}, \; d_2 = \frac{ln\left(\frac{F}{K}\right) - \frac{\sigma^2 T}{2}}{\sigma\sqrt{T}}$$

and the put option price can be derived similarly

$$P = Ke^{-rT}\phi(-d_2) - Fe^{-rT}\phi(-d_1)$$

## 10.2 Greeks - Solution

**Question :** Calculate the greeks $\Delta$, $\Gamma$, $\mathcal{V}$, $\rho$, $\Theta$ for a call option.

**Solution :** We denote $C$ the price of the call option at $t = 0$, $\phi$ the standard normal cumulative distribution and $f = \phi'$ the standard normal density function.

$$C = S_0\phi(d_1) - Ke^{-rT}\phi(d_2)$$

$$d_1 = \frac{ln\left(\frac{S_0}{K}\right) + rT + \frac{\sigma^2 T}{2}}{\sigma\sqrt{T}}, \quad d_2 = \frac{ln\left(\frac{S_0}{K}\right) + rT - \frac{\sigma^2 T}{2}}{\sigma\sqrt{T}} = d_1 - \sigma\sqrt{T}$$

We start with an identity that will help us for all the greeks

$$f(d_2) = \frac{1}{\sqrt{2\pi}}\exp\left(\frac{-d_2^2}{2}\right) = \frac{1}{\sqrt{2\pi}}\exp\left(\frac{-d_1^2}{2}\right)\exp\left(d_1\sigma\sqrt{T}\right)\exp\left(\frac{-\sigma^2 T}{2}\right)$$

$$f(d_2) = \frac{1}{\sqrt{2\pi}}\exp\left(\frac{-d_1^2}{2}\right)\frac{S_0}{K}\exp(rT) = f(d_1)\frac{S_0}{K}e^{rT} \tag{4}$$

- $\Delta$

$$\Delta = \frac{\partial C}{\partial S_0} = \phi(d_1) + S_0 f(d_1)\frac{\partial d_1}{\partial S_0} - Ke^{-rT}f(d_2)\frac{\partial d_2}{\partial S_0}$$

$$\Delta = \phi(d_1) + f(d_1)\frac{1}{\sigma\sqrt{T}} - Ke^{-rT}f(d_2)\frac{1}{S_0\sigma\sqrt{T}}$$

And using the equation 4

$$\Delta = \phi(d_1)$$

- $\Gamma$

$$\Gamma = f(d_1)\frac{1}{S_0\sigma\sqrt{T}}$$

- $\mathcal{V}$

$$\mathcal{V} = \frac{\partial C}{\partial \sigma} = S_0 f(d_1)\frac{\partial d_1}{\partial \sigma} - Ke^{-rT}f(d_2)\frac{\partial d_2}{\partial \sigma}$$

$$\mathcal{V} = \frac{\partial C}{\partial \sigma} = S_0 f(d_1)\frac{\partial d_1}{\partial \sigma} - Ke^{-rT}f(d_2)\left(\frac{\partial d_1}{\partial \sigma} - \sqrt{T}\right)$$

And using equation 4

$$\mathcal{V} = Ke^{-rT}f(d_2)\sqrt{T} = S_0 f(d_1)\sqrt{T}$$

- $\rho$

$$\rho = \frac{\partial C}{\partial r} = S_0 f(d_1)\frac{\partial d_1}{\partial r} + TKe^{-rT}\phi(d_2) - Ke^{-rT}f(d_2)\frac{\partial d_2}{\partial r}$$

$$\rho = S_0 f(d_1)\frac{\sqrt{T}}{\sigma} + TKe^{-rT}\phi(d_2) - Ke^{-rT}f(d_2)\frac{\sqrt{T}}{\sigma}$$

And using equation 4

$$\rho = TKe^{-rT}\phi(d_2)$$

- $\Theta$

$$\Theta = \frac{\partial C}{\partial T} = S_0 f(d_1)\frac{\partial d_1}{\partial T} + rKe^{-rT}\phi(d_2) - Ke^{-rT}f(d_2)\frac{\partial d_2}{\partial T}$$

$$\Theta = S_0 f(d_1)\frac{\partial d_1}{\partial T} + rKe^{-rT}\phi(d_2) - Ke^{-rT}f(d_2)\left(\frac{\partial d_1}{\partial T} - \frac{\sigma}{2\sqrt{T}}\right)$$

And using equation 4

$$\Theta = rKe^{-rT}\phi(d_2) + Ke^{-rT}f(d_2)\frac{\sigma}{2\sqrt{T}}$$

## 10.3 Ornsetin Uhlenbeck - Solution

**Question :** Derive the formula of an Ornstein Uhlenbeck process. Calculate its expectation and variance.

**Solution :** We start with the dynamics of an Ornsetin Uhlenbeck process

$$\mathrm{d}r_t = \theta\left(\mu - r_t\right)\mathrm{d}t + \sigma \mathrm{d}W_t$$

We consider the process $X_t = e^{At}r_t$. We apply Itô

$$\mathrm{d}X_t = Ae^{At}r_t\mathrm{d}t + e^{At}\mathrm{d}r_t$$

We find that the value $A = \theta$ gives

$$\mathrm{d}X_t = e^{\theta t}\left(\theta\mu\mathrm{d}t + \sigma\mathrm{d}W_t\right)$$

$$X_T - X_0 = \int_0^T \theta\mu e^{\theta t}\mathrm{d}t + \int_0^T \sigma e^{\theta t}\mathrm{d}W_t$$

$$r_T e^{\theta T} = r_0 + \mu(e^{\theta T} - 1) + \int_0^T \sigma e^{\theta t}\mathrm{d}W_t$$

$$r_T = r_0 e^{-\theta T} + \mu(1 - e^{-\theta T}) + \int_0^T \sigma e^{\theta(t-T)}\mathrm{d}W_t$$

We find that the expectation is

$$\mathbb{E}(r_T) = r_0 e^{-\theta T} + \mu(1 - e^{-\theta T})$$

and

$$\lim_{T\to\infty} \mathbb{E}(r_T) = \mu$$

We eliminate the deterministic terms for the variance, we find that

$$\text{Var}(r_T) = \text{Var}\left(\int_0^T \sigma e^{\theta(t-T)} \mathrm{d}W_t\right)$$

$$\text{Var}(r_T) = \mathbb{E}\left(\left(\int_0^T \sigma e^{\theta(t-T)} \mathrm{d}W_t\right)^2\right)$$

We apply the Itô isometry

$$\text{Var}(r_T) = \int_0^T \sigma^2 e^{2\theta(t-T)} \mathrm{d}t$$

$$\text{Var}(r_T) = \frac{\sigma^2}{2\theta}(1 - e^{-2\theta T})$$

and

$$\lim_{T\to\infty} \text{Var}(r_T) = \frac{\sigma^2}{2\theta}$$

## 10.4 Hybrid Vasicek - Solution

**Question :** Derive the relationship between the stock volatility and the rates volatility in a hybrid Vasicek model.

**Solution :** The interviewer is asking you to derive the classic calibration formula for equity model with stochastic rates. We consider the stock dynamics

$$\frac{\mathrm{d}S_t}{S_t} = r_t \mathrm{d}t + \sigma_t^S \mathrm{d}W_t^S$$

where $r_t$ follows

$$\mathrm{d}r_t = (\theta_t - \kappa r_t)\,\mathrm{d}t + \sigma_t^r \mathrm{d}W_t^r$$

Using the Ornstein Uhlenbeck derivation in the previous question we have

$$r_s = \int_t^s \exp(\kappa(u-s))\sigma_u^r \mathrm{d}W_u + \text{nonstochastic terms}$$

Therefore $S_T$ can be written

$$S_T = S_t P(t,T) exp\left(-\int_t^T \frac{\sigma_u^{s2} u}{2} \mathrm{d}u + \int_t^T \sigma_u^s W_u^s \mathrm{d}u\right)$$

where

$$P(t,T) = \mathbb{E}^{\mathrm{P}}\left[\exp\left(-\int_t^T r_s \mathrm{d}s\right)\right]$$
$$= \mathbb{E}^{\mathrm{P}}\left[\exp\left(-\int_t^T \int_t^s \exp(\kappa(u-s))\sigma_u^r \mathrm{d}W_u^r \mathrm{d}s\right)\right]$$

In order to process this integral we notice that

$$\int_t^T \int_t^s F(u,s)\mathrm{d}u\mathrm{d}s = \int_t^T \int_s^T F(u,s)\mathrm{d}s\mathrm{d}u$$

When applied in this case we obtain

$$P(t,T) = \mathbb{E}^{\mathrm{P}}\left[\exp\left(-\int_t^T \hat{B}(\kappa,u,T)\sigma_u^r \mathrm{d}W_u^r\right)\right]$$

where

$$\hat{B}(\kappa,u,T) = \frac{1-\exp(\kappa(u-T))}{\kappa}$$

We have found the volatility of the zero-coupon bond. We can find the drift using the fact that $P(t,T)$ is tradable, so $P(t,T)/B_t$ must be a $\mathbb{P}$-martingale and so we have

$$\frac{\mathrm{d}P(t,T)}{P(t,T)} = r_t\mathrm{d}t - \hat{B}(\kappa,t,T)\sigma_t^r \mathrm{d}W_t^r$$

Since $S_t$ is tradable, $S_t/P(t,T)$ will be a $\mathbb{Q}_T$-martingale, so it follows that

$$\frac{\mathrm{d}\left(\frac{X_t}{P(t,T)}\right)}{\frac{X_t}{P(t,T)}} = \sigma_t^S \mathrm{d}\widetilde{W}_t^S + \hat{B}(\kappa,t,T)\sigma_t^r \mathrm{d}\widetilde{W}_t^r$$

where $\tilde{W}_t$ are Brownian motions in $\mathbb{Q}_T$. It follows that under $\mathbb{Q}_T$, $S_T$ is log-normally distributed with mean $S_0/P(0,T)$ and variance

$$V_T = \int_0^T \left(\left(\sigma_t^S\right)^2 + 2\rho\sigma_t^S \hat{B}(\kappa,t,T)\sigma_t^r + \hat{B}(\kappa,t,T)^2\left(\sigma_t^r\right)^2\right)\mathrm{d}t$$

Now we calibrate the hybrid model to the market equity implied volatility by ensuring that

$$\sigma_{\mathrm{imp}}^2(T) = \frac{1}{T}\int_0^T \left(\left(\sigma_t^S\right)^2 + 2\rho\sigma_t^S \hat{B}(\kappa,t,T)\sigma_t^r + \hat{B}(\kappa,t,T)^2\left(\sigma_t^r\right)^2\right)\mathrm{d}t$$

Calibrating here means deriving $\sigma_t^S$. $\sigma_{\mathrm{imp}}$ is the market implied volatility (deduced from equity vanilla options prices) and $\sigma_t^r$ is the rates market implied volatility.

## 10.5 Fokker-Planck - Solution

**Question :** Derive the Fokker-Planck formula.

**Solution :**

Start with the SDE defined by

$$dX_t = \mu(X_t)\,dt + \sigma(X_t)\,dW_t$$

the transition density $\rho(x,t|y,s)$ is defined by

$$\begin{aligned}\int_A \rho(x,t|y,s)dx &= \Pr\left[X_{t+s} \in A | X_s = y\right] \\ &= \Pr\left[X_t \in A | X_0 = y\right]\end{aligned}$$

Consider a differentiable function $V(X_t,t) = V(x,t)$ with $V(X_t,t) = 0$ for $t \notin (0,T)$. Then by Itô's Lemma

$$dV = \left[\frac{\partial V}{\partial t} + \mu\frac{\partial V}{\partial x} + \frac{1}{2}\sigma^2\frac{\partial^2 V}{\partial x^2}\right]dt + \left[\sigma\frac{\partial V}{\partial x}\right]dW_t$$

so that

$$V(X_T,T) - V(X_0,0) = \int_0^T \left[\frac{\partial V}{\partial t} + \mu\frac{\partial V}{\partial x} + \frac{1}{2}\sigma^2\frac{\partial^2 V}{\partial x^2}\right]dt + \int_0^T \left[\sigma\frac{\partial V}{\partial x}\right]dW_t \quad (5)$$

where $\mu = \mu(X_t)$ and $\sigma = \sigma(X_t)$ for notational convenience. Take the conditional expectation of both sides of equation (5) given $X_0$

$$\begin{aligned}&E\left[V(X_T,T) - V(X_0,0)\right] \\ &= E\int_0^T \left[\frac{\partial V}{\partial t} + \mu\frac{\partial V}{\partial x} + \frac{1}{2}\sigma^2\frac{\partial^2 V}{\partial x^2}\right]dt + E\int_0^T \left[\sigma\frac{\partial V}{\partial x}\right]dW_t \quad (6) \\ &= \int_{\mathbb{R}}\left\{\int_0^T \left[\frac{\partial V}{\partial t} + \mu\frac{\partial V}{\partial x} + \frac{1}{2}\sigma^2\frac{\partial^2 V}{\partial x^2}\right]dt\right\}\rho(x,t \mid y,s)dx\end{aligned}$$

all expectations are expectations conditional on $X_0$, so that $E[\cdot] = E[\cdot \mid X_0 = y]$. since $E[dW_t] = 0$, the second term in the middle line of equation (6) drops out. Hence, we can write equation (6) as three integrals

$$\int_{\mathbb{R}}\int_0^T \rho\frac{\partial V}{\partial t}dtdx + \int_{\mathbb{R}}\int_0^T \rho\mu\frac{\partial V}{\partial x}dtdx + \frac{1}{2}\int_{\mathbb{R}}\int_0^T \rho\sigma^2\frac{\partial^2 V}{\partial x^2}dtdx = I_1 + I_2 + I_3 \quad (7)$$

where $\rho = \rho(x,t \mid y,s)$ for notational convenience. The objective of the derivation is to apply integration by parts to get rid of the derivatives of $V$. The trick is that

$I_1$ is evaluated using integration by parts on $t$, while $I_2$ and $I_3$ are each evaluated using integration by parts on $x$.
Use $u = \rho$, $v' = \frac{\partial V}{\partial t}$ so that $u' = \frac{\partial \rho}{\partial t}$ and $v = V$. Hence for the inside integrand of $I_1$ we have

$$\int_0^T \rho \frac{\partial V}{\partial t} dt = \rho V|_0^T - \int_0^T \frac{\partial \rho}{\partial t} V dt = -\int_0^T \frac{\partial \rho}{\partial t} V dt$$

since at the boundaries 0 and $T$, $V = 0$. Hence

$$I_1 = -\int_{\mathbb{R}} \int_0^T \frac{\partial \rho}{\partial t} V(x,t) dt dx$$

Change the order of integration in $I_2$ and write it as

$$I_2 = \int_0^T \int_{\mathbb{R}} \rho\mu \frac{\partial V}{\partial x} dx dt$$

Use integration by parts on the integrand, with $u = \rho\mu$, $v' = \frac{\partial V}{\partial x}$ so that $u' = \frac{\partial(\rho\mu)}{\partial x}$, $v = V$

$$\int_{\mathbb{R}} \rho\mu \frac{\partial V}{\partial x} dx = \rho\mu V|_{\mathbb{R}} - \int_{\mathbb{R}} \frac{\partial(\rho\mu)}{\partial x} V dx$$

Hence the integral can be evaluated as

$$\begin{aligned} I_2 &= -\int_0^T \int_{\mathbb{R}} \frac{\partial(\rho\mu)}{\partial x} V(x,t) dx dt \\ &= -\int_{\mathbb{R}} \int_0^T \frac{\partial(\rho\mu)}{\partial x} V(x,t) dt dx \end{aligned}$$

Finally, the evaluation of the integrand of $I_3$ requires the application of integration by parts on $x$ twice. This is because in the integrand we want to get rid of the $\frac{\partial^2 V}{\partial x^2}$ term and end up with $V(x,t)$ only. Again, change the order of integration and write $I_3$ as

$$\frac{1}{2} \int_0^T \int_{\mathbb{R}} \rho\sigma^2 \frac{\partial^2 V}{\partial x^2} dx dt$$

For the first integration by parts use $u = \rho\sigma^2$, $v' = \frac{\partial^2 V}{\partial x^2}$ so that $u' = \frac{\partial(\rho\sigma^2)}{\partial x}$ and $v = \frac{\partial V}{\partial x}$. Hence the integrand can be written

$$\begin{aligned} \int_{\mathbb{R}} \rho\sigma^2 \frac{\partial^2 V}{\partial x^2} dx &= \rho\sigma^2 \frac{\partial V}{\partial x}\bigg|_{\mathbb{R}} - \int_{\mathbb{R}} \frac{\partial(\rho\sigma^2)}{\partial x} \frac{\partial V}{\partial x} dx \\ &= -\int_{\mathbb{R}} \frac{\partial(\rho\sigma^2)}{\partial x} \frac{\partial V}{\partial x} dx \end{aligned}$$

Apply integration by parts again, with $u = \frac{\partial(\rho\sigma^2)}{\partial x}$, $v' = \frac{\partial V}{\partial x}$, $u' = \frac{\partial^2(\rho\sigma^2)}{\partial x^2}$, $v = V$

$$-\int_{\mathbb{R}} \frac{\partial(\rho\sigma^2)}{\partial x} \frac{\partial V}{\partial x} dx = -\frac{\partial(\rho\sigma^2)}{\partial x} V\bigg|_{\mathbb{R}} + \int_{\mathbb{R}} \frac{\partial^2(\rho\sigma^2)}{\partial x^2} V dx$$

$$= \int_{\mathbb{R}} \frac{\partial^2 (\rho\sigma^2)}{\partial x^2} V(x,t) dx$$

This implies that $I_3$ can be written as

$$\frac{1}{2} \int_0^T \int_{\mathbb{R}} \frac{\partial^2 (\rho\sigma^2)}{\partial x^2} V dx dt = \frac{1}{2} \int_{\mathbb{R}} \int_0^T \frac{\partial^2 (\rho\sigma^2)}{\partial x^2} V(x,t) dt dx$$

We can substitute $I_1$, $I_2$ and $I_3$ in (7)

$$E\left[V\left(X_T,T\right)\right] - V\left(X_0,0\right) = \int_{\mathbb{R}} \int_0^T V(x,t) \left[ -\frac{\partial \rho}{\partial t} - \frac{\partial(\rho\mu)}{\partial x} + \frac{1}{2}\frac{\partial^2 (\rho\sigma^2)}{\partial x^2} \right] \mathrm{d}t\mathrm{d}x$$

Since $V\left(X_t,t\right) = 0$ for $t \notin (0,T)$ we have $V\left(X_T,T\right) = V\left(X_0,0\right) = 0$ so that $E\left[V\left(X_T,T\right)\right] - V\left(X_0\right) = 0$. This implies that the portion of the integrand in the brackets is zero

$$-\frac{\partial \rho}{\partial t} - \frac{\partial(\rho\mu)}{\partial x} + \frac{1}{2}\frac{\partial^2 (\rho\sigma^2)}{\partial x^2} = 0$$

from which the Fokker-Planck equation can be obtained

$$\frac{\partial \rho}{\partial t} = -\frac{\partial(\rho\mu)}{\partial x} + \frac{1}{2}\frac{\partial^2 (\rho\sigma^2)}{\partial x^2}$$

## 10.6 Breeden-Litzenberger - Solution

**Question :** Derive the Breeden-Litzenberger Formula.

**Solution :** The Breeden-Litzenberger formula connects the underlying distribution to the derivatives of call options with respect to the strike. We start with the call price expectation formula

$$C(S,K,T) = e^{-rT}\mathbb{E}\left((S_T - K)^+\right) = e^{-rT} \int_K^\infty (x-k)p(x)\mathrm{d}x$$

where $p(x)$ is the density function of $S_T$. We take the derivative with respect to $K$ using the Leibniz formula (see page 128)

$$\frac{\partial C}{\partial K} = e^{-rT} \int_K^\infty -p(x)\mathrm{d}x$$

We take the second derivative to get the Breeden-Litzenberger formula.

$$\frac{\partial^2 C}{\partial K^2} = e^{-rT} p(k)$$

## 10.7 Local Volatility - Solution

**Question :** Derive the Dupire Formula or Local Volatility.

**Solution :** It might be surprising but this question is sometimes asked in interview. Local volatility is such an industry standard that you absolutely need to know it after a few years as a quant. Remember that the local volatility $\sigma_{loc}(t, S_t)$ when injected in the process $S_t$

$$\frac{dS_t}{S_t} = (r(t) - q(t))dt + \sigma_{loc}(t, S_t)dW_t$$

matches the vanilla prices, or in other terms, the implied volatility surface, because option prices are actually the implied volatility surface. Dupire proved the existence and uniqueness of the local volatility surface in his seminal 1994 paper. The idea is to connect the stock dynamics to the stock distribution using the Fokker Planck equation, and then to the call prices using the Breeden Litzenberger formula.

So we start by assuming that the stock prices follow

$$\frac{dS_t}{S_t} = (r(t) - q(t))dt + \sigma(t, S_t)dW_t$$

So the Breeden-Litzenberger formula between times $s$ and $T$, for a stock of spot value $s$, at time $t$, and discounting factor $D(t,T)$ (typically $D(t,T) = e^{r(T-t)}$) is

$$p(s, K, t, T) = \frac{1}{D(t,T)} \frac{\partial^2}{\partial K^2} C_t(s, K, T)$$

We apply Fokker-Planck to the stock

$$\tfrac{1}{2}\frac{\partial^2}{\partial x^2}\Big[\sigma(t,x)^2 x^2 p(x_0, x, t_0, t)\Big] - (r(t) - q(t))\frac{\partial}{\partial x}\Big[x p(x_0, x, t_0, t)\Big]$$

$$-\frac{\partial}{\partial t} p(x_0, x, t_0, t) = 0$$

We multiply this by $D(t_0, t)(x - K)^+$ and integrate from $x = K$ to $x = \infty$ to get

$$\begin{aligned}
&\tfrac{1}{2} D(t_0, t) \int_K^\infty \frac{\partial^2}{\partial x^2}\Big[\sigma(t,x)^2 x^2 p(x_0, x, t_0, t)\Big](x - K)dx \\
&-(r(t) - q(t))D(t_0, t) \int_K^\infty \frac{\partial}{\partial x}\Big[x p(x_0, x, t_0, t)\Big](x - K)dx \\
&-D(t_0, t) \int_K^\infty \frac{\partial}{\partial t} p(x_0, x, t_0, t)(x - K) = 0
\end{aligned} \tag{8}$$

The first term in equation (8) can be integrated by parts and using the Breeden-Litzenberger formula it can be rewritten as

$$\tfrac{1}{2}\sigma(t,K)^2K^2\frac{\partial^2}{\partial K^2}C_{t_0}(x_0,K,t) \tag{9}$$

The second term in equation (8) can be integrated by parts and using the integral version of the Breeden-Litzenberger formula it becomes

$$(r(t)-q(t))\left(C_{t_0}(x_0,K,t)-K\frac{\partial}{\partial K}C_{t_0}(x_0,K,t)\right) \tag{10}$$

The last term in equation (8) can be integrated directly and using

$$D_t(t_0,t)=-r(t)D(t_0,t)$$

we obtain

$$-\left(r(t)C(x,K,t)+\frac{\partial}{\partial t}C_{t_0}(x_0,K,t)\right) \tag{11}$$

After subtituting the three terms and rearranging equation (8) we obtain the Dupire Formula

$$\sigma(t,K)^2=\frac{(r(t)-q(t))K\frac{\partial}{\partial K}C_{t_0}(x_0,K,t)+\frac{\partial}{\partial t}C_{t_0}(x_0,K,t)+q(t)C_{t_0}(x_0,K,t)}{\frac{1}{2}K^2\frac{\partial^2}{\partial K^2}C_{t_0}(x_0,K,t)}$$

Usually we find it written in this compact form

$$\sigma_{\text{loc}}(t,K)^2=\frac{\frac{\partial C}{\partial t}+(r(t)-q(t))K\frac{\partial C}{\partial K}+q(t)C}{\frac{1}{2}K^2\frac{\partial^2 C}{\partial K^2}}=\frac{C_t+(r-q)KC_k+qC}{\frac{1}{2}K^2C_{KK}}$$

## 10.8 Black Scholes Equation - Solution

**Question :** Derive the Black Scholes equation.

**Solution :** We consider a financial option on a stock $S_t$ of value $V(S_t,t)$. We construct a self-financed portfolio with the option and a $\Delta$ hedging amount of stock.

$$P=V(S,t)+\Delta S$$

The stock has the usual dynamics

$$\frac{\mathrm{d}S_t}{S_t}=r\mathrm{d}t+\sigma\mathrm{d}W_t$$

The self-financing portfolio condition is

$$dP=dV+\Delta dS \tag{12}$$

And the non arbitrage condition gives us

$$dP = rPdt \tag{13}$$

We apply Itô on $V$, and combining (12) and (13) we have

$$V_t dt + V_S \mathrm{d}S + \frac{1}{2} V_{SS} \mathrm{d}\langle S \rangle_t + \Delta \mathrm{d}S = rP\mathrm{d}t$$

Therefore we have the hedging amount $\Delta$

$$\Delta = -V_S$$

and

$$V_t dt + \frac{1}{2} V_{SS} \mathrm{d}\langle S \rangle_t = rV\mathrm{d}t - rSV_S \mathrm{d}t$$

We use the stock dynamics and simplify the d$t$

$$V_t + \frac{1}{2} V_{SS} \sigma^2 S^2 = rV - rSV_S$$

We obtain the Black Scholes equation, usually written

$$V_t + \frac{1}{2} V_{SS} \sigma^2 S^2 + rSV_S - rV = 0$$

It is sometimes found in its forward version

$$V_t + \frac{1}{2} V_{FF} \sigma^2 F^2 - rV = 0$$

which is derived using the derivation rule

$$\frac{\partial V(F,t)}{\partial t} = \frac{\partial V}{\partial t} + \frac{\partial F}{\partial t}\frac{\partial V}{\partial F} = V_t + rFV_F$$

## 10.9 Black Scholes Robustness - Solution

**Question :** You sell a European option for which you estimated the volatility to be $\sigma_t$. What is the PnL error if you hedge this option until expiry and that the realized volatility turns out to be $\sigma_r \neq \sigma_t$?

**Solution :** This classic result was first discussed by El Karoui, Jeanblanc and Shreve (1996) and Carr and Madan (1997) and is known as the PnL tracking error formula or Black Scholes robustness formula. Let $V$ be the price of the option. It is designed at inception to verify the Black Scholes equation with $\sigma_t$

$$V_t + \frac{1}{2} V_{SS} \sigma_t^2 S^2 + rSV_S - rV = 0 \tag{14}$$

On another hand, we apply Itô's lemma to $V$, which evolves during the life of the trade at the realized volatility $\sigma_r$

$$dV = V_t dt + V_S dS + \frac{1}{2} V_{SS} \sigma_r^2 S^2 dt \tag{15}$$

We isolate $V_t$ in (14), substitute it in (15) and integrate from 0 to $T$

$$V(T) - V(0) = \int_0^T rV dt - \int_0^T rSV_S dt + \int_0^T V_S dS + \int_0^T \frac{1}{2} V_{SS} (\sigma_r - \sigma_t)^2 S^2 dt \tag{16}$$

We see that the 2 first integrals correspond to the cost of holding a position in cash equal to $(V - SV_S)$, which is the position in cash after selling the option at $V$ and hedging it with a $V_S$ position in stock

$$I_1 = \int_0^T rV dt - rSV_S dt$$

The third integral is the change in value of a $V_S$ position in stock

$$I_2 = \int_0^T V_S dS$$

And we find that the difference in value of the european claim is equal to its expected hedging cost $(I_1 + I_2)$ plus an error term

$$\text{PnL}_{\text{error}} = \int_0^T \frac{1}{2} V_{SS} (\sigma_r - \sigma_t)^2 S^2 dt$$

## 10.10 Local Variance as an Expectation - Solution

**Question :** Show that the local variance can be seen as a conditional expectation of the instantaneous variance.

**Solution :** This result was derived by Derman and Kani (1998). Following the derivation by Derman and Kani, assume the stock follows the usual dynamics, and let us consider the forward price process

$$F_{t,T} = S_t e^{\int_t^T \mu_s ds}$$

$$\frac{dF_{t,t}}{F_{t,T}} = \sqrt{v_t} dW_t$$

and

$$dF_{T,T} = dS_T$$

The undiscounted value of a European option with strike $K$ expiring at time $T$ is given by

$$C(S_0, K, T) = \mathbb{E}\left[(S_T - K)^+\right]$$

Differentiating once with respect to $K$ gives

$$\frac{\partial C}{\partial K} = -\mathbb{E}\left[\theta(S_T - K)\right]$$

where $\theta(\cdot)$ is the Heaviside function. Differentiating again with respect to $K$ gives

$$\frac{\partial^2 C}{\partial K^2} = \mathbb{E}\left[\delta(S_T - K)\right]$$

where $\delta(\cdot)$ is the Dirac $\delta$ function. Now a formal application of Itô's lemma to the terminal payoff of the option (and using $dF_{T,T} = dS_T$) gives the identity

$$d(S_T - K)^+ = \theta(S_T - K)\, dS_T + \frac{1}{2} v_T S_T^2 \delta(S_T - K)\, dT$$

Taking conditional expectations of each side, and using the fact that $F_{t,T}$ is a martingale, we get

$$dC = d\mathbb{E}\left[(S_T - K)^+\right] = \frac{1}{2}\mathbb{E}\left[v_T S_T^2 \delta(S_T - K)\right] dT$$

Also, we can write

$$\begin{aligned}\mathbb{E}\left[v_T S_T^2 \delta(S_T - K)\right] &= \mathbb{E}\left[v_T \mid S_T = K\right] \frac{1}{2} K^2 \mathbb{E}\left[\delta(S_T - K)\right] \\ &= \mathbb{E}\left[v_T \mid S_T = K\right] \frac{1}{2} K^2 \frac{\partial^2 C}{\partial K^2}\end{aligned}$$

Putting this together, we get

$$\frac{\partial C}{\partial T} = \mathbb{E}\left[v_T \mid S_T = K\right] \frac{1}{2} K^2 \frac{\partial^2 C}{\partial K^2}$$

Comparing this with the forward version of local volatility (see page 126) we see that

$$\sigma^2(K, T, S_0) = \mathbb{E}\left[v_T \mid S_T = K\right]$$

That is, local variance is the risk-neutral expectation of the instantaneous variance conditional on the final stock price $S_T$ being equal to the strike price $K$.

Chapter 11

# Math Cheatsheet

# Math Cheatsheet

## 11.1 Normal Distribution

$$X \sim \mathcal{N}\left(\mu, \sigma^2\right)$$

$$f_X(x) = \frac{1}{\sqrt{2\pi}\sigma} \exp\left(-\frac{(x-\mu)^2}{2\sigma^2}\right)$$

## 11.2 Correlation

The population correlation coefficient $\rho_{X,Y}$ between two random variables $X$ and $Y$ with standard deviations $\sigma_X$ and $\sigma_Y$ is defined as

$$\rho_{X,Y} = \frac{\operatorname{cov}(X,Y)}{\sigma_X \sigma_Y}$$

## 11.3 Brownian Motion

A Brownian motion is a stochastic process $\{B_t\}_{t \geq 0+}$ with the following properties:

- $B_0 = 0$
- The function $t \to B_t$ is almost surely continuous in $t$
- The process $\{B_t\}_{t \geq 0}$ has stationary, independent increments
- The increment $B_{t+s} - B_s$ has the $\mathcal{N}(0, t)$ distribution

## 11.4 $\sigma$-algebra

Let $\Omega$ be a set. A collection $\mathcal{A}$ of subsets of $\Omega$ is a $\sigma$-algebra on $\Omega$, if and only if it satisfies all the following properties:

- $\Omega, \emptyset \in \mathcal{A}$
- For all $A \in \mathcal{A}, A^c \in \mathcal{A}$
- For all sequence $(A_n)_{n=1}^{\infty}$ of elements of $\mathcal{A}, \cup_{n=1}^{\infty} A_n \in \mathcal{A}$

## 11.5 Martingale

An $(\mathcal{F}_t)$-adapted, real-valued process $M$ is called a martingale (with respect to the filtration $(\mathcal{F}_t)$ if

- $\mathrm{E}\,|M_t| < \infty$ for all $t \in T$
- $\mathrm{E}\,(M_t|\mathcal{F}_s) \overset{\text{as.}}{=} M_s$ for all $s \leq t$

## 11.6 Girsanov

Let $B_t, 0 \leq t \leq T$ be a Brownian motion on a probability space $(\Omega, \mathcal{F}, P)$, and let $\mathcal{F}_t, 0 \leq t \leq T$, be a filtration for this Brownian motion. Let $a_t$ be an adapted process. Define

$$Z_t = \exp\left(-\int_0^t a_u dB_u - \frac{1}{2}\int_0^t a_u^2 du\right)$$

$$\tilde{B}_t = B_t + \int_0^t a_u du$$

and the probability $\tilde{P}$ equivalent to $P$ defined by

$$\tilde{P}(A) = \int_A Z(\omega)dP(\omega)$$

and assume that

$$E\left[\int_0^t a_u^2 du\right] < +\infty$$

Then under the probability $\tilde{P}$ the process $\tilde{B}$ is a Brownian motion.

## 11.7 Itô Process

A process $X_t$ is said to be an Itô process if there exist progressively measurable processes $\alpha_t$ and $\beta_t$ such that

$$\int_0^t \left(|\alpha_s| + \beta_s^2\right) ds < \infty, \text{ a.s.}$$

$$X_t = X_0 + \int_0^t \alpha_S ds + \int_0^t \beta_s dB_s$$

## 11.8 Itô's Lemma

Let $f(t, x)$ be a real-valued function whose second-order partial derivatives are continuous. Let $(X_t)_{t\geq 0}$ be an Itô process, Then

$$df(t, X) = \frac{\partial f}{\partial t}dt + \frac{\partial f}{\partial x}dX + \frac{1}{2}\frac{\partial^2 f}{\partial x^2}d\langle X\rangle_t$$

In practice, it is more convenient to use the notation

$$df = f_t dt + f_X dX + \frac{1}{2} f_{XX} d\langle X\rangle_t$$

## 11.9 Levy Theorem

Let $M_t$ be a martingale with continuous sample paths and $M_0 = 0$. Then

$$d\langle M\rangle_t = dt \iff M \text{ is a Brownian motion}$$

## 11.10 Martingale representation Theorem

Let $B_t$ be a Brownian motion and $\mathcal{F}_t$ the augmented filtration generated by $B_t$. If $X$ is an $\mathcal{F}_\infty$-measurable square integrable random variable, then there is a unique $\mathcal{F}_t$-adapted predictable process $\phi$, such that

$$X = \mathbb{E}[X] + \int_0^\infty \phi_s dB_s$$

## 11.11 Doob's Optional Sampling Theorem

If M is a martingale and $S, T$ are stopping times with

$$S \le T \text{ a.s. and } \mathbb{E}\,|M_T| < +\infty$$

then

$$\mathbb{E}\left[M_T|\mathcal{F}_S\right] = M_S$$

## 11.12 Doob's Optional-Stopping Theorem

Let $(\Omega, \Sigma, \mathbf{P})$ be a probability space, $\mathcal{F} = \{F_n\}$ a filtration on $\Omega$, and $X = \{X_n\}$ a martingale with respect to $\mathcal{F}$. Let $\tau$ be a stopping time. Suppose that any one of the following conditions holds:

- There is a positive integer $N$ such that $\tau(\omega) \le N$ for all $\omega \in \Omega$
- There is a positive real number $K$ such that

  $$|X_n(\omega)| < K$$

  for all $n$ and all $\omega \in \Omega$, and $\tau$ is almost surely finite.
- $\mathbb{E}(\tau) < \infty$, and there is a positive real number $K$ such that

  $$|X_n(\omega) - X_{n-1}(\omega)| < K$$

  for all $n$ and all $\omega \in \Omega$ Then $X_T$ is integrable, and

  $$\mathbb{E}\left(X_\tau\right) = \mathbb{E}\left(X_0\right)$$

## 11.13 Doob's Optional-Stopping Theorem (continuous)

Let $(M_t)_{t\geq 0}$ be a right continuous $\mathcal{F}_t$ -martingale and $\tau$ be a stopping time with respect to $\mathcal{F}_t$. If either one of the following conditions holds:

- $\tau$ is bounded, i.e. $\exists N < \infty$ such that $\tau \leq N$;
- $\exists c > 0$ such that $E\left[|M_t|\right] \leq c, \forall t > 0$,

then $E\left[M_\tau\right] = E\left[M_0\right].$

## 11.14 Long-term behavior of trajectories

Let $\{B_t\}_{t\in[0,\infty)}$ be a Brownian motion. Then,

$$\limsup_{t\to\infty} \frac{B_t}{\sqrt{t}} = \infty, \ a.s$$

and

$$\liminf_{t\to\infty} \frac{B_t}{\sqrt{t}} = -\infty, \ a.s$$

## 11.15 Stopping Time

*A* (generalized) random variable $T$ is called a stopping time if $T : \Omega \longrightarrow \mathbb{Z}_+ \cup \{\infty\}$ satisfies $\{T \leq n\} \in \mathcal{F}_n$.

## 11.16 Hitting Times (First Passage Times)

Let $T_a = \min\{t : B(t) = a\}$ be the first time the standard Brownian motion process hits a. Using the reflection principle we can prove that

$$\mathrm{P}\left(T_a \leq t\right) = 2\mathrm{P}(B(t) \geq a) = 2 - 2\Phi(a/\sqrt{t})$$

$$\lim_{t\to\infty} \mathrm{P}\left(T_a \leq t\right) = 1$$

## 11.17 Itô's Isometry

Let $B_t$ be a Brownian Motion and $X_t$ a stochastic process

$$\mathrm{E}\left[\left(\int_0^T X_t \mathrm{d}B_t\right)^2\right] = \mathrm{E}\left[\int_0^T X_t^2 \mathrm{d}t\right]$$

## 11.18 Bayes' theorem

$$P(A|B) = \frac{P(B|A)P(A)}{P(B)} = \frac{P(A \cap B)}{P(B)}$$

where $A$ and $B$ are events and $P(B) \neq 0$

## 11.19 Self-financing Portfolios

A portfolio, or trading strategy, is any predictable process

$$\phi = (\phi_0, \dots, \phi_n)$$

Its corresponding value process is

$$V(t) = V(t; \phi) := \sum_{i=0}^{n} \phi_i(t) S_i(t)$$

The portfolio $\phi$ is called self-financing (for $S$) if the stochastic integrals

$$\int_0^t \phi_i(u) dS_i(u), \quad i = 0, \dots, n$$

are well defined and

$$dV(t; \phi) = \sum_{i=0}^{n} \phi_i(t) dS_i(t)$$

## 11.20 Uniform Integrability Test Functions

A function $\psi : [0, \infty) \to [0, \infty)$ is called a u.i. (uniform integrability) test function if $\psi$ is increasing, convex (i.e. $\psi(\lambda x + (1-\lambda)y) \leq \lambda\psi(x) + (1-\lambda)\psi(y)$ for all $x, y \in [0, \infty), \lambda \in [0, 1]$) and

$$\lim_{x \to \infty} \frac{\psi(x)}{x} = \infty$$

So for example $\psi(x) = x^p$ is a u.i. test function if $p > 1$.

## 11.21 Uniform Integrability Theorem

The family $\{f_j\}_{j \in J}$ is uniformly integrable if and only if there is a u.i. test function $\psi$ such that

$$\sup_{j \in J} \left\{ \int \psi\left(|f_j|\right) dP \right\} < \infty$$

## 11.22 Doob's martingale convergence theorem

Let $N_t$ be a right-continuous supermartingale. Then the following are equivalent:

1. $\{N_t\}_{t \geq 0}$ is uniformly integrable
2. There exists $N \in L^1(P)$ such that $N_t \to N$ a.e. $(P)$ and $N_t \to N$ in $L^1(P)$, i.e. $\int |N_t - N| \, dP \to 0$ as $t \to \infty$

## 11.23 Local Volatility

The local volatility is defined as

$$\sigma_{\text{loc}}(t, K)^2 = \frac{C_t + (r-q)KC_k + qC}{\frac{1}{2}K^2 C_{KK}}$$

If we express the option price as a function of the forward price

$$F_t = S_0 e^{\int_0^T \mu_t dt}$$

where $\mu_t$ represents the risk-neutral drift of the stock process, we get the forward version of local volatility

$$\sigma_{\text{loc}}(t, K)^2 = \frac{C_t}{\frac{1}{2}K^2 C_{KK}}$$

## 11.24 Breeden-Litzenberger

First order form

$$\frac{\partial C}{\partial K} = e^{-rT} \int_K^{\infty} -p(x) \mathrm{d}x$$

Second order form

$$\frac{\partial^2 C}{\partial K^2} = e^{-rT} p(k)$$

## 11.25 Fokker-Planck

$$\frac{\partial \rho}{\partial t} = -\frac{\partial(\rho\mu)}{\partial x} + \frac{1}{2}\frac{\partial^2 (\rho\sigma^2)}{\partial x^2}$$

## 11.26 Vanilla Options

Call option

$$C = S_0 e^{-qT} \phi(d_1) - K e^{-rT} \phi(d_2)$$

$$d_1 = \frac{ln\left(\frac{S_0}{K}\right) + (r-q)T + \frac{\sigma^2 T}{2}}{\sigma\sqrt{T}}, \quad d_2 = \frac{ln\left(\frac{S_0}{K}\right) + (r-q)T - \frac{\sigma^2 T}{2}}{\sigma\sqrt{T}}$$

Put option

$$P = Ke^{-rT}\phi(-d_2) - S_0e^{-qT}\phi(-d_1)$$

## 11.27 Reflection Principle

Let $W_t$ a Brownian motion, and $a > 0$, then the reflection principle states:

$$\mathbb{P}\left(\sup_{0\leq s\leq t} W(s) \geq a\right) = 2\mathbb{P}(W(t) \geq a)$$

## 11.28 Tanaka's Formula

$$|B_t| = \int_0^t \operatorname{sgn}(B_s)\, dB_s + L_t$$

where $B_t$ is the standard Brownian motion, sgn denotes the sign function and $L_t$ is its local time at 0 (the local time spent by $B$ at 0 before time t) given by the $L_2$-limit

$$L_t = \lim_{\varepsilon\downarrow 0} \frac{1}{2\varepsilon}|\{s \in [0,t]|B_s \in (-\varepsilon, +\varepsilon)\}|$$

## 11.29 Symmetric Matrices

Any symmetric matrix $A$ ($A = A^T$)

- has only real eigenvalues
- is always diagonalizable
- has orthogonal eigenvectors

## 11.30 Semidefinite Positive Matrices

The symmetric matrix $A$ is said positive semidefinite ($A \geq 0$) if all its eigenvalues are non negative.

## 11.31 Useful Taylor Series

$$
\begin{aligned}
\frac{1}{1-x} &= 1+x+x^2+x^3+x^4+\dots \\
\left(\frac{1}{1-x}\right)' = \frac{1}{(1-x)^2} &= 1+2x+3x^2+4x^3+5x^4+\dots \\
e^x &= 1+x+\frac{x^2}{2!}+\frac{x^3}{3!}+\frac{x^4}{4!}+\dots \\
\cos x &= 1-\frac{x^2}{2!}+\frac{x^4}{4!}-\frac{x^6}{6!}+\frac{x^8}{8!}-\dots \\
\sin x &= x-\frac{x^3}{3!}+\frac{x^5}{5!}-\frac{x^7}{7!}+\frac{x^9}{9!}-\dots
\end{aligned}
$$

## 11.32 Leibniz Integral Rule

Let $f(x,t)$ be a function such that both $f(x,t)$ and its partial derivative $f_x(x,t)$ are continuous in $t$ and $x$ in some region of the $(x,t)$-plane, including $a(x) \le t \le b(x)$, $x_0 \le x \le 1$. Also suppose that the functions $a(x)$ and $b(x)$ are both continuous and both have continuous derivatives for $x_0 \le x \le x_1$. Then, for $x_0 \le x \le x_1$,

$$\frac{d}{dx}\left(\int_{a(x)}^{b(x)} f(x,t)dt\right) = f\left(x, b(x)\right) b'(x) - f(x, a(x))a'(x) + \int_{a(x)}^{b(x)} \frac{\partial}{\partial x} f(x,t)dt$$

# Index

Made in the USA
Monee, IL
08 January 2025